Designing Big Data Platforms

Designing Big Data Platforms

How to Use, Deploy, and Maintain Big Data Systems

Yusuf Aytas
Dublin, Ireland

Registered Office

John Wiley & Sons, Inc., 111 River Street, Hoboken, NJ 07030, USA

Editorial Office

111 River Street, Hoboken, NJ 07030, USA

For details of our global editorial offices, customer services, and more information about Wiley products visit us at www.wiley.com.

Library of Congress Cataloging-in-Publication Data Applied for:

ISBN: 9781119690924

Cover design by Wiley
Cover image: © monsitj/iStock/Getty Images

Set in 9.5/12.5pt STIXTwoText by Straive, Chennai, India

10 9 8 7 6 5 4 3 2 1

Contents

List of Contributors

Ömer E. Avşaroğulları
San Francisco, CA
United States

Louis Calisi
Acton, MA
United States

Ender Demirkaya
Seattle, WA
United States

Alperen Eraslan
Ankara
Turkey

Ang Gao
Dublin
Ireland

Zehra Kavasoğlu
London
United Kingdom

David Kjerrumgaard
Henderson, Nevada
United States

Ari Miller
Los Angeles, CA
United States

Alper Mermer
Manchester
United Kingdom

Will Schneider
Waltham, MA
United States

Preface

Big data has been one of the buzzwords of the last couple of decades. Although there is a certain hype around big data, it is now seen as increasingly essential for organizations. There are so many new systems popping up now and then. This book does not particularly focus on any of the new systems or projects, but rather focuses on how these systems are built and used. The book gives an overview of the systems and lays out the architecture. The book mentions and sometimes goes into detail about both new and old big data systems, such as Apache Hadoop and Apache Spark; it attempts to give an overall view over designing big data platforms regardless of the technology.

Who Should Read this Book

The book offers some general knowledge about designing big data platforms. If you are a front-end engineer, backend engineer, data engineer, data analyst, or data scientist, you will see different aspects of designing big data platforms. The book at times goes into technical detail on some subjects through code or design but this doesn't prevent the non-specialist obtaining an understanding of the underlying concepts. If you are an expert on big data platforms, perhaps, the book can revisit things of which you might already be aware.

Scope of this Book

The book gives a general overview of big data technologies to design big data platforms. The book covers many interesting technologies, but it is not a reference book for any of the technologies mentioned. It dives deep into certain technologies but overall tries to establish a framework rather than focusing on certain tools.

Outline of this Book

At the beginning, the book tries to go over big data, big data platforms, and a simple data processing system. Later, it starts to go into discussing various aspects of big data such as storage, processing, discovery, security, and so forth. At the end of the book, it summarizes systems that it talked about and discusses some of the useful patterns for designing big data platforms. In Appendix A, the book discusses some of the other technologies and patterns that don't really fit the flow of the book. In Appendix B, the book discusses recipes where it presents a solution for a particular big data problem.

5 February 2021   *Yusuf Aytas*
Dublin

Acknowledgments

This book is an overview of many systems, patterns, and ideas, in an attempt to combine experiences from academia and industry with reference to many great articles, posts, and books in order to provide a consolidated view on designing big data platforms. While writing this book, I have gained enormous knowledge from many exceptional authors, engineers, analysts, scientists, and editors. I'm very grateful for all the information that people made available.

Yusuf Aytas

Acronyms

ACL	access control list
API	application programming interface
CI	continues integration
CPU	central processing unit
CQL	Cassandra Query Language
CSV	comma separated values
DAG	directed acyclic graph
DBMS	data base management system
DB	database
DDD	domain driven design
DS	date string
ETL	extract, transform, load
FIFO	first in first out
GPU	graphical processing unit
HDFS	Hadoop distributed file system
HH	hour string
HQL	hive query language
HTML	hypertext markup language
IO	input/output
IP	Internet Protocol
JSON	JavaScript Object Notation
LB	load balancer
ML	machine learning
NoSQL	not only SQL
OS	operating system
RDD	resilient distributed dataset
SDK	software development kit
SLA	service level agreement
SLO	service level objective

SQL	structured query language
TCP	transport control protocol
TH	threshold
URL	uniform resource locator
VM	virtual machine
XML	extensible markup language
YARN	yet another resource manager

Introduction

The value of big data and processing data to get actionable insights have been a competitive advantage across the world. Organizations are still in a rush to collect, process, and drive values from big data. Designing big data platforms to collect, store, process, and discover big data is a significant challenge from many aspects. Thanks to collaborative push from engineers from different parts of the world and several organizations, we have so many great systems we can use to design a big data platform.

A big data platform consists of many components where we have many alternatives for the same job. Our task is to design a platform that caters to the needs and requirements of the organization. In doing so, we should choose the right tool for the job. Ideally, the platform should adapt, accept, and evolve due to new expectations. The challenge is to design a simple platform while keeping it cost-efficient in terms of development, maintenance, deployment, and the actual running expense.

In this book, I present many different technologies for the same job. Some of them are already off the shelf, while others are cutting edge. I want to give perspectives on these systems so that we can create solutions that are based on the experience of others. Hopefully, we can all have a better grasp on designing big data platforms after reading this book.

1

An Introduction: What's a Modern Big Data Platform

After reading this chapter, you should be able to:

- Define a modern Big Data platform
- Describe expectations from data
- Describe expectations from a platform

This chapter discusses the different aspects of designing Big Data platforms, in order to define what makes a big platform and to set expectations for these platforms.

1.1   Defining Modern Big Data Platform

The key factor in defining Big Data platform is the extent of data. Big Data platforms involve large amounts of data that cannot be processed or stored by a few nodes. Thus, Big Data platform is defined here as an infrastructure layer that can serve and process large amounts of data that require many nodes. The requirements of the workload shape the number of nodes required for the job. For example, some workloads require tens of nodes for a few hours or fewer nodes for days of work. The nature of the workloads depends on the use case.

Organizations use Big Data platforms for business intelligence, data analytics, and data science, among others, because they identify, extract, and forecast information based on the collected data, thus aiding companies to make informed decisions, improve their strategies, and evaluate parts of their business. The more the data recorded in different aspects of business, the better the understanding. The solutions for Big Data processing vary based on the company strategy.

Companies can either use on-site or cloud-based solutions for their Big Data computing and storage needs. In either case, various parts can be considered all

Designing Big Data Platforms: How to Use, Deploy, and Maintain Big Data Systems,
First Edition. Yusuf Aytas.

together as a Big Data platform. The cogs of the platform might differ in terms of storage type, compute power, and life span. Nevertheless, the platform as a whole remains responsible for business needs.

1.2 Fundamentals of a Modern Big Data Platform

What makes a modern Big Data platform remains unclear. A modern Big Data platform has several requirements, and to meet them correctly, expectations with regard to data should be set. Once a base is established for expectations from data, we can then reason about a modern platform that can serve it.

1.2.1 Expectations from Data

Big Data may be structured, semi-structured, or unstructured in a modern Big Data platform and come from various sources with different frequencies or volumes. A modern Big Data platform should accept each data source in the current formats and process them according to a set of rules. After processing, the prepared data should meet the following expectations.

1.2.1.1 Ease of Access

Accessing prepared data depends on internal customer groups. The users of the platform can have a very diverse set of technical abilities. Some of them are engineers, who would like to get very deep and technical with the platform. On the other hand, some may be less technically savvy. The Big Data platform should ideally serve both ends of the customer spectrum.

Engineers dealing with the platform expect to have an application programming interface (API) to communicate with about the platform in various integration points. Some of the tasks would require coding or automation from their end. Moreover, data analysts expect to access the data through standard tooling like SQL or write an extract, transform, load (ETL) job to extract or analyze information. Lastly, the platform should offer a graphical user interface to those who simply want to see a performance metric or a business insight even without a technical background.

1.2.1.2 Security

Data is an invaluable asset for organizations. Securing data has become a crucial aspect of a modern Big Data platform. Safeguarding against a possible data breach is a big concern because a leak would result in financial loses, reduced customer trust, and damage to the overall reputation of the company.

Security risks should be eliminated, but users should be able to leverage the platform easily. Achieving both user-friendliness and data protection requires a combination of different security measures such as authentication, access control, and encryption.

The organizations should identify who can access to the platform. At the same time, access to a particular class of data should be restricted to a certain user or user group. Furthermore, some of the data might contain critical information like PII, which should be encrypted.

1.2.1.3 Quality

High-quality data enables businesses to make healthier decisions, opens up new opportunities, and provides a competitive advantage. The data quality depends on factors such as accuracy, consistency, reliability, and visibility. A modern Big Data platform should support ways to accomplish accurate and consistent data between data sources to produce visible data definition and reliable processed data. The domain is the driving factor for a Big Data platform when it comes to data quality. Hence, the number of resources allocated to the data quality changes according to the domain. Some of the business domains might be quite flexible, while others would require strict rules or regulations.

1.2.1.4 Extensibility

Iterative development is an essential part of software engineering. It is no surprise that it is also part of Big Data processing. A modern Big Data platform should empower the ease of reprocessing. Once the data is produced, the platform should provide infrastructure to extend the data easily. This is an important aspect because there are many ways things can go wrong when dealing with data. One or more iteration can be necessary.

Moreover, the previously obtained results should be reproducible. The platform should reprocess the data and achieve the same results when the given parameters are the same. It is also important to mention that the platform should offer mechanisms to detect deviations from the expected result.

1.2.2 Expectations from Platform

After establishing expectations regarding the data, how to meet these expectations by the platform should be discussed. Before starting, the importance of the human factor should be noted. Ideal tooling can be built, but these would be useful only in a collaborative environment. Some of the critical business information and processing can occur with good communication and methods. This section will present an overview of the features in pursuit of our ideal Big Data platform; we will not go into detail in explaining each of the features we would employ since we have chapters discussing it.

1.2.2.1 Storage Layer

Ideally, a storage layer that can scale in terms of capacity, process an increasing number of reads and writes, accept different data types, and provide access permissions. Typical Big Data storage systems handle the capacity problem by scaling horizontally. New nodes can be introduced transparently to the applications backed by the system. With the advent of cloud providers, one can also employ cloud storage to deal with the growing amount of storage needs. Moreover, a hybrid solution is an option where the platform uses both on-site and cloud solutions. While providing scalability in terms of volume and velocity, the platform should also provide solutions in cases of backup, disaster recovery, and cleanups.

One of the hard problems of Big Data is backups as the vast amount of storage needed is overwhelming for backups. One of the options for backups is magnetic tapes as they are resilient to failures and do not require power when they are not in use. A practical option is relying on durable and low-cost cloud storage. In addition, an expensive but yet very fast solution is to have a secondary system that either holds partly or the whole data storage. With one of the proposed solutions in place, the platform can potentially perform periodic backups.

In case of disaster recovery from backups, separate sets of data sorted by their priority are an option since retrieving backup data would take quite some time. Having different data sets also provides the ability to spin up multiple clusters to process critical data in parallel. The clusters can be spun up on separate hardware or again using a cloud provider. The key is to be able to define which data sets are business-critical. Categorizing and assigning priority to each data set enables the recovery execution to be process-driven.

The storage layer can suffer from lost space when the data are replicated in many different ways but no process is available to clean up. There are two ways to deal with data clean up. The first is the retention policy. If all data sets have a retention policy, then one could build processes to flush expired data whenever it executes. The second is the proactive claiming of unused data space. To understand which data is not accessed, a process might look at the access logs and determine unused data. Hence, a reclaiming process should be initiated by warning the owners of the data. Once the owners approve, the process should be initiated and reclaim the space.

1.2.2.2 Resource Management

The workload management consists of managing resources across multiple requests, prioritization of tasks, meeting service-level agreements (SLAs), and assessing the cost. The platform should enable important tasks to finish on time, respond to ad hoc requests promptly, and use available resources judiciously to complete tasks quickly and measure the cost. To accomplish these, the platform should provide an approach for resource sharing, visibility for the entire platform, monitoring around individual tasks, and cost reporting structure.

Resource sharing strategies can affect the performance of the platform and fairness toward individual jobs. On one hand, when there is no task running, the platform should use as much resources as possible to perform a given task. On the other hand, a previously initiated job slows down all other requests that started after this task. Therefore, most of the Big Data systems provide a queuing mechanism to separate resources. Queuing enables sharing of resources across different business units. On the other hand, it is less dramatic when the platform uses cloud-based technologies. A cloud solution can give the platform the versatility to run tasks on short-lived clusters that can automatically scale to meet the demand. With this option, the platform can employ as many nodes as needed to perform tasks faster.

Oftentimes, the visibility of the platform in terms of usage might not be a priority. Thus, making a good judgment is difficult without easily accessible performance information. Furthermore, the platform can consist of a different set of clusters, which then makes it even harder to visualize the activity in the platform at a snapshot of time. For each of the technology used under the hoot, the platform should be able to access performance metrics or calculate itself and report them in multiple graphical dashboards.

The number of tasks performed on the platform slows down a cluster or even bring it down. It is important to set SLAs for each performed task and monitor individual tasks for their runtime or resource allocation. When there is an oddity in executing tasks, the platform should notify the owner of the task or abort the task entirely. If the platform makes use of cloud computing technologies, then it is extremely important to abort tasks or not even start executing them by using the estimated costs.

I believe the cost should be an integral part of the platform. It is extremely important to be transparent for the customers. If the platform can tell how much it cost or can cost to run their workloads, it would be customers of the platform to decide how much money they can spend. The team maintaining the platform would not be responsible for the cost. If one business unit wants to spin up a big cluster or buy new hardware, then it is their problem to justify the need.

1.2.2.3 ETL

ETL stands for extract, transform, and load. ETL is the core of Big Data processing; therefore, it is the heart of a modern Big Data platform. The Big Data platform should provide an ETL solution/s that manages the experience end to end. The platform should control the flow from data generation to processing and making means out of the data. ETL developers should be able to develop, test, stage, and deploy their changes. Besides, the platform should hide technical details where possible and provide advanced features.

The size of the company is a factor for the number of storage system required because this system should be able to support multiple sources and targets for a given ETL engine. The more integration points it offers, the more useful the ETL engine becomes. Ideally, the ETL engine should have the plug-in capability where each kind of data source/target is configured by additional plug-ins. When there is a demand for a new source/target, the platform would simply require another plug-in to support a new source/target.

The platform should encourage ownership of flows and data. The ETL engine should make it obvious underlying data is owned by the same user group. If a user does not have rights to modify the flow, the right to access the data is not granted, or vice versa. The ETL engine itself may require exclusive rights on the data storage layer to manage access permissions for user and user groups.

The support for the development life cycle is an important aspect. The platform should be able to let developers build their ETLs potentially locally, test the flow, review the changes, stage the changes, and finally deploy to production. The key to local development is the ability to generate partial test data. Since the platform should also accept the data, partial creation of source data should be made easy by supplying a sampling percent. Once the test data is available, testing becomes much easier.

In most of the modern ETL tooling, an intuitive user interface might be missing for the creation of flows. Common ETL engines require some understanding of technical details such as source control and column mapping. Some users of the platform may not be technically savvy or it might be genuinely easier to just deal with user interface rather than coding. A user interface to drag and drop data from multiple sources and merge in different ways would assist to configure trivial flows faster.

1.2.2.4 Discovery

The meaning of data can get lost quickly in big organizations. As the amount of the data grows, so does the metadata. To ensure that the metadata definition is shared across the company, the platform should offer metadata discovery capabilities. The metadata should be collected from various resources into a single repository where the definition of metadata can be updated or modified to reflect the context. Additional information such as owner, lineage, and related information would be useful when reviewing metadata. Moreover, the repository should be quickly searchable by various dimensions.

Nobody likes manual jobs. The platform should provide a data discovery tool that automatically updates metadata information by crawling each data source configured. When crawling for metadata, the discovery tool should get information such as attribute definition, type, and technology-specific information, e.g. partition key. Once the information is stored in a single repository, the relevant information should be shown to the users where they can update any information related to the metadata.

The discovery tool will use other information like queries or foreign keys to form the lineage where possible. Additional processing and storage will be necessary as most of the storage engines would not keep queries forever. If the queries are related to metadata, one can give a sample of queries when the metadata is viewed. Finding the owner of the data can be tricky since the owner of the table would not reveal much because the metadata may be from an actual team or group. Thus, ownership may be dealt with semi-automated fashion by having people from the organization confirm the group for the given set of metadata.

A single repository brings the ability to search for everything in one place. The search should have the ability to filter metadata information by type and data source. The metadata should be also searchable by any attribute or definition.

1.2.2.5 Reporting

Reporting is a necessary process for any business to quickly review the performance and status of the different areas of the business. A modern Big Data platform should provide a tool to present rich visualizations, user-friendly interface for exploring, dashboard creation/sharing, and aggregation functions on how data sources are displayed. Furthermore, the tooling should seamlessly integrate with the existing data storage layer.

The ability to show the data in a wide array of visualizations helps to quickly understand the summary. To make visualizations faster, the tooling will rely on client-side caching to avoid querying underlying data storage. This is a significant optimization as it both saves from computing power and gives the chance to swiftly load the requested dashboard.

Once reporting tooling supports common paradigms like SQL, it is easy to integrate most of the SQL supported storage engines. The tooling should support various drivers to communicate with the storage layer and retrieve data from various data sources. The tooling itself should understand the SQL to generate the query to load the dashboard and apply aggregation functions.

1.2.2.6 Monitoring

As it happens in any other platform, many systems or user errors would occur in a Big Data platform. An upgrade to the storage layer may change how currencies are handled or how someone can calculate the item price in euros incorrectly. On top of this, there might be node failures, update to connection settings, and more. All of these problems could delay or interrupt data processing. As the complexity of the system grows, so does the number of edge cases. Consequently, Big Data platforms are quite complex as they are built based on distributed systems. Preparation for failures is the only solution as even the detection of problems is complex. The platform should have protection against node/process failures, validation for schema and data, and SLAs per task or flow.

Nodes fail and processes lag. Even though most of the Big Data systems are designed to deal with occasional hiccups, failures still become problematic in practice. The Big Data platform should monitor the health of each Big Data system. The best way to verify everything, at least those that are functional, is to have small tasks executed against each of the systems. If these small tasks fail for one or more reasons, a manual intervention should be undertaken. If the problems could not be detected early, this would lead to the disruption of one or more data flows. Sometimes, the small tasks would be executed with no problem, but bigger tasks would lag due to various reasons. Such problems should be resolved on the flow level as big tasks could not be processed and especially the specific ones are causing errors. We should have SLAs for the delivery of full flow. If it does not meet the agreement, the problem should be escalated within the platform.

The platform should also check schema changes for data flowing through the systems. A schema validation framework is necessary to ensure that the changes to the schema are backward compatible. Moreover, validating schema itself is not enough. The data can be corrupted even if it conforms to validations. A new change may introduce the corruption of data at its core. To deal with such problems, basic anomaly detection should be performed and complex anomaly detection might be required. The basic anomaly detection would be only checking counts or number of duplicates, while a complex anomaly detection requires complex queries over time. The platform should offer both solutions as protection mechanisms.

1.2.2.7 Testing

Ideally, each part of the Big Data platform should have relevant test suits. However, testing is often skipped at many stages due to the pressure of the expected productivity but with errors. Other than decent unit test coverage, the platform should perform integration tests between the systems, performance tests, failure testing, and automation for running tests.

The importance of isolating a component and ensuring it produces the expected behavior of the given input is undisputable. Yet, under implementation, such suits might seem somewhat cumbersome for Big Data platforms. One reason is the need for stubbing for various external systems when testing. However, we cannot ensure any other way than unit testing to verify the component behaves as we expected. Thus, it is necessary to have unit tests in place to continuously validate the behavior against new changes.

Big Data platforms have many systems underneath. Each system has different ways to communicate. Additionally, these systems need to talk to each other. Sometimes an upgrade or a new change might break the contract. To detect such issues before they make it to production, we should have integration test suits between the systems. The integration test suits should ideally run for every change that is pushed to any of the systems. If running per change is difficult, then the integration tests can be scheduled to run multiple times a day to detect potential issues.

The load testing aspect is crucial when a new system gets introduced to a Big Data platform. Since we are working with Big Data systems, a virtual load should be created in a staging environment by streaming the expected volume of data to the new system. The expected volume can be estimated by a series of prediction analysis. Once the volume is confirmed, the data should be fed and the system validated that it can cope with it. Moreover, we should also stress the system with extra load. We would like to answer questions about the best throughput vs latency in different scenarios or the point where the system gets unresponsive.

Occasionally, testing the system with extreme scenarios is a beneficial exercise to see the worst-case scenario. Users may want to see how the system behaves in the presence of a kernel panic in multiple nodes or in a split-brain scenario. Moreover, it is interesting to monitor where the system experiences CPU slowdown, high packet loss, and slow disk access. One can add many other exercises to test against. Lastly, we would like to see how the system degrades with random problems.

1.2.2.8 Lifecycle Management

Designing, developing, and maintaining Big Data systems is complicated and requires all-around team effort and coordination. We need to draw the big picture and coordinate teams or team members according to the needs. Otherwise, we would end up in situations where nobody knows what to do next or get lost in rabbit holes. To prevent such frustrations, a structured plan is needed, where the progress for each component is visible and the next steps are clear. Hence, I propose a common structure with the phases as follows: planning, designing, developing, maintenance, and deprecation.

The planning phase involves resource allocation, cost estimation, and schedule of the Big Data systems. In the design phase, the requirements are met and a prototype is built. Once we have a working system, integrations and interactions are designed with other systems or end users. The next phase is development, where the software are built and the deployment pipelines including several test suits are prepared. Once the software is ready, the maintenance phase begins. If for some reason we decide not to invest in the system, we would go to the deprecation phase where our clients/customers will be moved from the system to the alternative offer.

2

A Bird's Eye View on Big Data

After reading this chapter, you should be able to

- Learn chronological information about Big Data processing
- List qualities that characterize Big Data
- List components of Big Data platforms
- Describe uses cases for Big Data

Development of Big Data platforms has spanned for over two decades. To provide an overview of its evolution, this chapter will present the qualities of Big Data, components of Big Data platform, and use cases of Big Data.

2.1 A Bit of History

Computing has advanced drastically in the past two decades, from network to data storage, with significant improvements. Despite the rapid changes, the definition of Big Data remains relevant. This section presents the evolution of Big Data chronologically.

2.1.1 Early Uses of Big Data Term

The term Big Data was used and described by Cox and Ellsworth (1997), who presented two ideas: Big Data collections and Big Data objects. Big Data collections are streamed by remote sensors as well as satellites. The challenge is pretty similar to today's Big Data where data is unstructured and has different data sources. Big Data objects are produced from large-scale simulations of computational dynamics and weather modeling. The combined problem of the Big Data object

Designing Big Data Platforms: How to Use, Deploy, and Maintain Big Data Systems,
First Edition. Yusuf Aytas.
© 2021 John Wiley & Sons, Inc. Published 2021 by John Wiley & Sons, Inc.

and Big Data collections are again comparable to today's Big Data challenges where data is too large for memory and disk to fit a single machine.

In his presentations regarding Big Data, Mashey (1998) noted that the need for storage has been growing faster and more data are being created in the Internet. Given the explosion of widely accessible data lead to problems as regards creating, understanding, and moving it, Mashey (1998) concluded that processing large amounts of data would require more computing, network, and disks and thus, more machines to distribute data.

At the time, Big Data had become popular. Weiss and Indurkhya (1998) reported that at the start of Big Data revolution, running data mining algorithms was similar to operating a warehouse and discussed the concepts related to extract, transform, load (ETL) for data mining purposes. Law et al. (1999) refers to the multi-threaded streaming pipeline architecture for large structured data sets to create visualization systems. The popularity of the term has been increasing even more since 2000 and cited in many academic articles such as Friedman et al. (2000), and Ang and Teo (2000), among others.

2.1.2   A New Era

The uses of Big Data were unknown up until Dean and Ghemawat (2004) introduced MapReduce, whose paradigm drastically shifted the perspective as regarding processing of Big Data. It is a simple yet very powerful programming model that can process large sets of data. The programmers specify a Map function that generates intermediary data that is fed into a Reduce function to subsequently merge values. The MapReduce program uses a set of input key/value pairs and produces a set of output key/value pairs. The programmer specifies two functions: Map and Reduce. The Map function takes the user input and produces intermediary output. The framework then shuffles the intermediary key/value pairs such that intermediary keys belong to the same node. Once shuffled, the Reduce function takes an intermediary key and set of values associated with the key and merges them into smaller values.

2.1.2.1   Word Count Problem

Let us see this powerful programming model in action. Consider the following code where we would count occurrences of a keyword in given documents.

```
//emit function simply writes key,value pairs
//map function
const map = (key, value) => {
  // key: document name
  // value: document contents
```

```
  for (const word of value.split(' ')) {
    emit(word, 1);
  }
};
//reduce function
const reduce = (key, values) => {
  // key: a word
  // values: a list of counts
  let sum = 0;
  for (const count of values) {
    sum += count;
  }
  emit(key, sum);
};
```

In the Map function, we simply take the document name and its contents. We iterate through each word and emit the result as a key(word)/value(count) pair. Afterward, the framework collects all pairs with the same key(word) from all outputs and groups them together. Grouped pairs then feed into the Reducer function that simply sums the count then emits the result as a key(word)/value(sum) pair.

2.1.2.2 Execution Steps
The Map function calls are partitioned across multiple machines by M splits. Once the mapping phase is complete, the intermediary keys are partitioned across machines by R pieces using a Partition function. The list of actions occur is illustrated in Figure 2.1.

1. The MapReduce framework splits the inputs into M pieces of typically block size and then begins running copies of the program on a cluster of machines. One copy of the program becomes the master copy that is used when assigning tasks to other machines.
2. In the mapping phase, machines pull the data locally, run the Map function, and simply emit the result as intermediary key/value pairs. The intermediary pairs are then partitioned into R partitions by the Partition function.
3. The MapReduce framework then shuffles and sorts the data by the intermediary key.
4. When a reduce slave has read all intermediate data, it then runs the Reduce function and outputs the final data.

The MapReduce framework is expected to handle very large data sets so the master keeps track of slaves and checks out every slave periodically. If it cannot receive a response back from the slave for a given timeout period, it then marks the task as failed and schedules the task on another machine. The rescheduling of

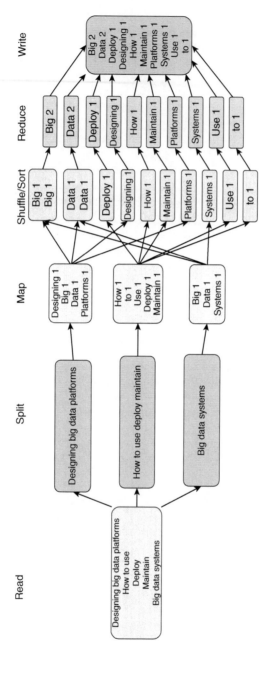

Figure 2.1 MapReduce execution steps.

tasks is possible since each task is idempotent. In the case of master failure, the program can be started from the last checkpoint.

2.1.3 An Open-Source Alternative

Shortly after Google published the paper on MapReduce, facing the same issues while developing Apache Apache Nutch (2004), a web crawler, Doug Cutting decided to use this new programming model. He switched the infrastructure behind the web crawler to his new implementation of MapReduce, Hadoop, named after his son's elephant toy (Olson, 2010). As being a direct competitor to Google, Yahoo! decided to support the development of Hadoop. Hadoop became the open-source solution for the Big Data explosion.

The use of weblogs which recorded the activity of the user on the website along with structured data became a valuable source of information for companies. As there was no commercially available software to process large sets of data, Hadoop became the tool for processing Big Data. Hadoop enabled companies to use commodity hardware to run jobs over Big Data. Instead of relying on the hardware to deliver high availability, Hadoop assumes all hardware are prone to failure and handles these failures automatically. It consists of two major components: a distributed file system and a framework to process jobs in a distributed fashion.

2.1.3.1 Hadoop Distributed File System

The Apache Hadoop (2006) Distributed File System (HDFS) is a fault-tolerant, massively scalable, a distributed file system designed to run on commodity hardware. HDFS aims to deliver the following promises:

- Failure recovery
- Stream data access
- Support very large data sets
- Write once and read many times
- Collocate the computation with data
- Portability

NameNode and DataNodes HDFS is based on master/slave architecture where NameNode is the master and DataNodes are slaves. NameNode coordinates the communication with clients and organizes access to files by clients. NameNode manages file system namespace and executes operations like opening, closing, renaming, and deleting files and determines the mapping of the data over DataNodes. On the other hand, DataNodes are many and live in the nodes with data and simply serve read/write requests from the file system clients as well as for instructions by the NameNode.

File System HDFS supports traditional file system organization. A client can create a directory and store multiple files or directories in this directory. NameNode keeps the information about file system namespace. Any update to the file system has to go through the NameNode. HDFS supports user quotas and access permissions. Moreover, HDFS allows clients to specify the replication count per file.

NameNode keeps the entire file system properties including Blockmap which contains blocks to files in memory. For any update, it logs every transaction to a transaction log called EditLog. A new file creation or replication factor change results in an entry to EditLog. Moreover, NameNode saves the in-memory data to a file called FsImage and truncates the EditLog periodically. This process is called a checkpoint. The period for creating checkpoints is configurable. When NameNode restarts, it reads everything from FsImage and applies additional changes that are recorded in EditLog. Once the transactions are safely written, NameNode can truncate the EditLog and create another checkpoint.

Data Replication HDFS can store very large files over a cluster of machines. HDFS divides each file into blocks and replicates them over machines by the replication factor. Files are written once and read many times. So, modification to files is not supported except for appending and truncation. The NameNode keeps track of all the blocks for a given file. Moreover, NameNode periodically receives a block report from the DataNodes that contains the list of data blocks that DataNode has.

The placement of data over machines is critical for HDFS's reliability. HDFS employs a rack aware replication policy to improve the resiliency of the system against node and rack failures. For the common use case of a replica count of 3, HDFS writes a replica of the block to a random node in the same rack and another replica of the block to another node in another rack. This replication behavior is illustrated in Figure 2.2.

Handling Failures HDFS has to be resilient to failures. For this reason, DataNodes send heartbeat messages to NameNode. If NameNode does not receive the heartbeat message due to node failure or a network partition, it marks these DataNodes as failed and the death of a DataNode can cause the number of replicas for a block to become less than the minimum. NameNode tracks these events and initiates block replication when necessary.

HDFS can occasionally move data from one DataNode to another when the space for DataNode decreases below a certain threshold. Moreover, HDFS can decide to replicate more if there is a fast-growing demand on a certain file. When interacting with DataNodes, the client calculates a checksum of files received and retrieves another copy of the block if the file is corrupt.

NameNode is the single point of failure for HDFS. If the NameNode fails, manual intervention is needed. To prevent catastrophic failures, HDFS has the option to keep FsImage and EditLog in multiple files. A common approach to deal with

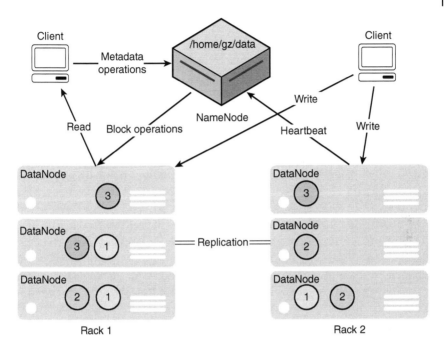

Figure 2.2 HDFS architecture.

this is to write FsImage and EditLog to highly available storage like NFS or distributed edit log.

Data Organization As HDFS stores files in blocks, applications write their data to these blocks once and read them repeatedly. When a client begins writing data to HDFS, NameNode finds the list of DataNodes where the client will write the data. The client then starts writing data to the first DataNode. After that, the DataNode starts replicating the data to the next one on the list, and then the next on the list starts replicating to the third one the list. HDFS pipelines data replication to any number of replicas where each one receives data and replicates it to the next node on the list.

The writes from the client do not arrive at a DataNode directly. The client stages the data worthy of HDFS block size and contacts NameNode for the DataNode to write. After that, the client writes to the specified destination. When the write operation ends, the client contact NameNode and NameNode commits file creation operation.

2.1.3.2 HadoopMapReduce
HadoopMapReduce is an open-source implementation of Google's MapReduce programming model. HadoopMapReduce runs compute nodes next to DataNodes

to allow the framework to execute tasks in the same node the data lives. The MapReduce framework executes MapReduce jobs with the help of YARN. YARN, yet another resource negotiator, is a job scheduling and resource management system to cater to the needs of the Big Data.

The rising adoption of MapReduce led to new applications of Big Data due to its availability in HDFS and multiple possible ways to process the same data. In addition, MapReduce is batch-oriented, hence missing the support for real-time applications. Having an existing Hadoop cluster to do more is cost-effective in terms of administration and maintenance (Murthy et al., 2014). Hence, the Hadoop community wanted a real multitenancy solution for these requirements.

YARN Architecture Separating resource management from job scheduling and monitoring is a primary design concern for YARN. YARN provides a resource manager and application master per application. The resource manager is the arbitrator to share resources among applications and employs a node manager per node to control containers and monitor resource usage. On the other hand, the application manager negotiates resources with the resource manager and works alongside the node manager to execute jobs.

The resource manager has two major components: scheduler and applications manager. The scheduler is responsible for allocating resources for applications and a pure scheduler as it is just limited to perform scheduling tasks; the scheduler neither tracks nor monitors nor start failed tasks. The scheduler schedules containers that are an abstraction over CPU, network, disk, memory, and so forth. The scheduler has a pluggable policy for allocating resources among applications based on their queues.

ApplicationsManager accepts job submissions, negotiates for Application-Master containers, monitors them, and restarts these containers upon failure. ApplicationMaster negotiates resources with the scheduler and keeps track of the containers.

As shown in Figure 2.3, this architecture allows YARN to scale better because there is no bottleneck at the resource manager. The resource manager does not have to deal with the monitoring of the various applications. Furthermore, it also enables moving all application-specific code to the application master so that one can perform other tasks than MapReduce. Nevertheless, YARN has to protect itself from ApplicationMasters since it is the user code.

YARN Resource Model YARN provides a pretty generic resource model. The YARN resource manager can track any countable resources. By default, it monitors CPU and memory for all applications and queues. YARN is designed to handle multiple applications at once. To the so, the scheduler has extensive knowledge about

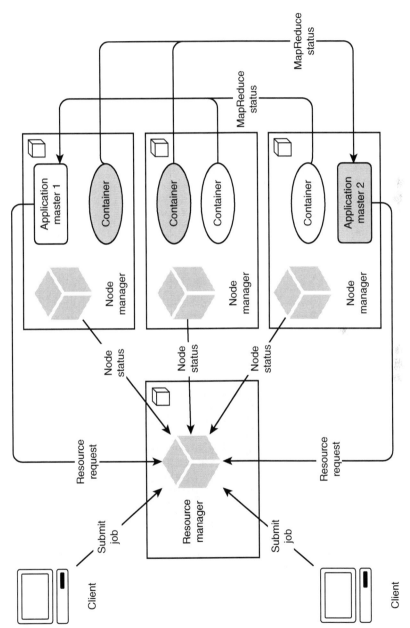

Figure 2.3 YARN architecture.

the status of the resources. Each ApplicationMaster negotiates with the resource manager for resources by a resource request with the following attributes:

- **Priority** of the request.
- **Resource name** of the host or rack on which the allocation is desired.
- **Resource** required for each request, e.g. CPU/memory.
- **Number of containers**, of above specifications, which are required by the application.
- **Relax locality**, defaulting to true, tells the ResourceManager if the application wants locality to be loose.

2.2 What Makes Big Data

Big Data has been described by many authors from diverse perspectives. Mayer-Schönberger and Cukier (2013) refers to Big Data as things one can do at a large scale that cannot be done at a smaller one, to extract new insights or create new forms of value, in ways that change markets, organizations, the relationship between citizens and governments, and more. Jacobs (2009) defines Big Data as the size of data forces one to look beyond the tried-and-true methods that are prevalent at that time, while for Pope et al. (2014), these are large and complex digital datasets that typically require nonstandard computational facilities for storage, management, and analysis. Russom et al. (2011) observed that Big Data has become a growing challenge that organizations face as they deal with large and fast-growing sources of data or information that also present a complex range of analysis and use problems. Lohr (2012) describes Big Data as a meme and a marketing term that is also a shorthand for advancing trends in technology that open the door to a new approach to understanding the world and making decisions. I think all of the above definitions describe Big Data quite well from different aspects. I define Big Data as a new breed of data, resulting from technological advancements in storage and a growing understanding of value extraction, different than traditional structured data in terms of volume, velocity, variety, and complexity.

2.2.1 Volume

The volume of the data is rapidly growing. Simply, the world generates more data on a year over year basis ranging from monitoring to logging. There are more applications, more devices, and more sensors. While the volume of data is getting bigger, it keeps being one of the predominant characteristics of Big Data.

What we think is large today may not be as large in the future since data storage technologies continue to advance. Moreover, the type of data also defines whether it is relatively big. If we store video files, then it is rather easy to get high volumes rapidly. If we collect activity logs, the storage need might not get as fast. So, defining "big" depends on the type of data and the amount of storage used.

While the common argument for defining what is big has been the number of machines to store the data, I argue the number of dedicated machines is the defining factor. This is a subtle detail as several virtual machines are available for storing data in the age of cloud computing. Moreover, the number of machines to make the data big is unclear and difficult to define, but I argue it should be at least in the order of 10s.

2.2.2 Velocity

The rate of data accumulated per second drastically increases every day owing to more sales transactions and more social activity per second. The number of new devices and sensors drives the need for real-time processing. Companies want to deliver the best experience to their users within a fraction of a second by making use of generated data. The real-time data provide valuable information about demographics, location, and past activities that companies use to deliver real customer value.

The push for real-time analytics creates the need for capturing and processing the data as fast as possible. Conventional data platforms are not suitable to deal with huge data feeds spontaneously. With the growing demand, we see the Big Data technologies becoming part of the puzzle. Hence, velocity is another defining factor for Big Data.

2.2.3 Variety

The Internet generates data from various sources ranging from social networks, database tables, spreadsheets, activity feeds, emails, a sensor source, or IoT devices for different activities from user tracking to activity feeds. The variety of data sources result in heterogeneous data (structured, semi-structured, and unstructured data). Most of these sources can be consumed readily and require preprocessing for integration into an application. To make holistic decisions, businesses want to use all the data that they can get.

A variety of data is not required, but several integration points for several applications of data. Nevertheless, the emergence of the need for processing new data sources makes it different. Thus, in Big Data, the variety is appreciated because it enables informed assumptions, predictions, and analysis.

2.2.4 Complexity

The complexity for Big Data is an unavoidable and perpetual challenge due to its other characteristics such as the variety and the fluctuating rate of incoming data and merging them. We might also need to get the Cartesian product of huge volumes. It does not stop here. Different data from disparate sources should be cleansed and filtered and each data type should be converted into a common format.

Though none of the above challenges are unique to Big Data, these are common. The value extraction process is complex, and automating it is another layer of complication. Despite this, businesses expect a continuous stream of processed data for informed decision-making. Thus, complexity becomes an accepted aspect of Big Data.

2.3 Components of Big Data Architecture

Big Data platforms may consist of many parts such as batch processing, real-time analytics, reporting, business intelligence, machine learning and so on. Based on the business needs, a Big Data platform may use some of the parts or all of them at the same time. For simplicity, we combine these parts into four main components: ingestion, storage, computation, and presentation as shown in Figure 2.4.

2.3.1 Ingestion

Data ingestion is the component where the platform accepts data from several sources and where ETL occurs. The data come in many different formats and thus require cleaning and filtering. The ingestion layer should have appropriate methods for dealing with each different format. Extraction involves retrieving and

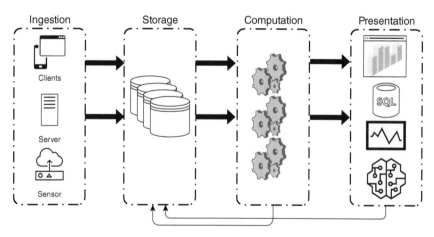

Figure 2.4 Components of Big Data architecture.

collecting data from source systems. Transformation requires executing several steps of business rules to convert source format to the destination. Data loading is simply inserting the data into the target. The key here is dealing with types between the source format and destination. Often one part might be too loose where the other part is strict. If the platform has an API to accept data, then it can also receive feeds of data from components within the platform.

2.3.2 Storage

The storage component is where we have our data warehouses or our data lakes. The storage layer can have multiple technologies to store the data since it is better to employ the right tool for the job. Moreover, organizations depending on the size might require more than one solution. Instead, different units within the company might have different requirements, e.g. storing graph data.

An important factor when it comes to Big Data is having a good judgment of the underlying system. Learning details of the storage technology helps in decision-making, understanding the limitations, and assessing it for new problem domains. For an overall idea, we might categorize systems by their consistency model, flexibility in the schema, or storage format.

It is common to see Big Data storage systems sacrifice consistency for the sake of availability or performance. It might be preferable to have eventual consistency instead of strong consistency when receiving high writes. When working with such systems, we have to operate through workarounds to achieve the results we want.

A common solution is to employ highly available and expandable distributed key-value data storage systems where we do not necessarily have a strong schema. This is a common way to allow flexibility in application development. Nevertheless, joining such data with others might become quickly very costly.

Another important factor can be choosing between row or column-oriented data. If the clients do not access all columns of the data, it might make sense to use columnar storage because one might save disk space, reduce I/O when retrieving data, and improve query execution time.

2.3.3 Computation

Computation component is the layer where we derive insights, predictions, models, and more from the data. We implement methods to analyze large data and find patterns. We combine and train valuable data buried in a bunch of disorganized datasets. In doing so, we make use of domain knowledge to get the best out of the computations.

Computation involves many areas such as clustering, correlation, and regression analysis. Moreover, we compute to get information like forecasting or machine learning models. Delivering results for all these computations might require different types of processing. Some of them might require offline processing of days worth of data. On the other hand, some of them require real-time processing. Sometimes, we might need both for the same problem domain because computing everything in real-time might be too costly.

2.3.4 Presentation

The presentation component of the architecture is where we make use of the computed data and make a further computation on the fly. This is what business cares for most of the time. Rightly so, this is the part where the business builds more engagement, better relationship, and personalized experience with their customers. Moreover, forecasting or analysis of the current trends is made readily available for exploring.

The tooling here involves data visualization tools and APIs as well as real-time machine learning models and services. The visualization tools harness the aggregated data and unleash the potential opportunities within the data. The models or API help delivering better and personalized customer experience. Moreover, other tooling helps in forecasting and various statistical analysis.

2.4 Making Use of Big Data

Big Data is becoming the standard facility for any company or organization. At this point, organizations from different branches of business or research widely adopted Big Data technologies. Depending on the area, some organizations use some of the methods of Big Data and sometimes all of them at once. Institutions make use of Big Data in the form querying, reporting, alerting, searching, exploring, mining, and modeling.

2.4.1 Querying

The ad hoc analysis is a big part of the formulation of new insights, building the right data sets, and understanding one or more metrics quickly. SQL is the most common way to query data, and it is no different for Big Data. Both open source and cloud engines execute SQL queries over large sets of data. The key is to make critical data sets available for querying.

2.4.2   Reporting

Once the post-processing of data is done, an effective reporting mechanism for different parts of organizations follows. Reports are sent to the executive team and upper management to understand the direction of the company. There are also reports that focus on a specific part of the business and those that are shared with third-party companies. Aggregated data sets where the truth and metrics lie are used in all these types of reports.

2.4.3   Alerting

Big Data enables monitoring of the business in multiple directions since the data come from various sources. An anomaly in any part of the business can be detected with an intelligent threshold and that anomalous part will be alerted. Such alerting can prevent losing customers or potential revenue, and such monitoring can be automated to be more efficient.

2.4.4   Searching

Big Data technologies empower fast and bulk indexing of the data. The indexing would take longer if the data should be brought into a destination. Nevertheless, the data are already available in Big Data systems. Having such data makes indexing easier and allows real-time indexing. Moreover, one can go even further to record top searches and optimize them. Saving search queries and processing them again requires Big Data technologies. Thus, Big Data creates a feedback loop that was not possible before.

2.4.5   Exploring

Due to the opportunities within Big Data brought by current technologies, exploration in the absence or limited information can lead to hypothesis generation. Once some ideas are formulated, then one can tie it to real indicators. Exploring a big set of data comes in handy since it offers a better world view.

2.4.6   Mining

Clustering, classification, collaborative filtering, and frequent pattern mining for various parts of the business may add a huge value when offering customers different services. Instead of showing unrelated advertisements or products, one can organize information in different ways that match customer profiles. Big Data helps to merge different sources and enabling new dimensions to understand user behavior. Moreover, the processing of data becomes more comfortable since the data can be poured into one place.

2.4.7 Modeling

Organizations aim to provide the best services on demand for their customers. They want to modify service behavior on-demand based on customer or user input. To do so, they train models offline over Big Data and deploy them as new services. Big Data facilitates developing machine learning models on-demand over dozens of data sets. What's more, it enables calculating multiple models with different inputs since the data can be processed in parallel without restriction. One can then use multiple models to observe A/B testing results on them to drive better outcomes.

3

A Minimal Data Processing and Management System

After reading this chapter, you should be able to:

- Use Unix tooling for data processing
- Build a pipeline to extract information
- Automate common data tasks
- Use PostgreSQL for large data

Overcomplicating a problem more than it has to be is a general issue in computing. It is hard to per se "keep it simple stupid." Experience and time are essential to formulate solutions that are good enough for the problem. In this chapter, solutions for relatively large data are developed using standard tools and open-source software.

3.1 Problem Definition

Defining a problem and working on this through different strategies are the focus in this chapter. Let us assume we are hosting an online book store.

3.1.1 Online Book Store

Imagine a scenario where we have an online book store. The online bookstore has a shopping process where consumers directly buy books from the seller in real time. The book store has an interface for searching the books and a landing page with some of the recommended books. As the user navigates through the website, he/she can add books to the shopping basket. At any time, the user can decide to buy books and check out. If the user is not registered, the website first attempts to register the user. After that, the user can choose a shipping address. Finally, the website offers payment options.

Designing Big Data Platforms: How to Use, Deploy, and Maintain Big Data Systems, First Edition. Yusuf Aytas.
© 2021 John Wiley & Sons, Inc. Published 2021 by John Wiley & Sons, Inc.

3.1.2 User Flow Optimization

A special attention is given to the user flow optimization. We aim to have a better understanding of user behavior throughout checkout. Once some log data are obtained, the user behavior can be analyzed, and website experience can be optimized in the areas that can be improved.

Let us say we have application logs that provide timestamp, host, IP, HTTP method, unique cookie per user, path, HTTP status code, content length, request time in milliseconds, HTTP referrer, and user agent. In reality, there might be many more things than this for logging purposes related to the request metrics. Nevertheless, we cut it short to make it simple:

```
2019-12-30:20:27:45.983 dbdp.web01.dev.com 212.129.72.174 GET
c2a436df-dec2-440e-9d67-dd9a9729f526/basket/checkout 200 16 10 None
Mozilla/5.0 (Macintosh; Intel Mac OS X 10.14; rv:72.0)Gecko/20100101 Firefox/72.0
```

3.2 Processing Large Data with Linux Commands

In this section, a problem is presented that requires access to logs files from a bunch of servers to reason about a certain problem. Standard Linux commands and additional helper commands are used to process the data in different directions.

3.2.1 Understand the Data

One of the first things to do is to understand the data. We previously explained what log files contained. However, this might not be readily available, so we have to go to the source to acquire related information. Linking to the source file is suggested where the logs are generated when writing an application that parses logs.

After studying the data, some parameters are considered that might be helpful and filter the rest of the data. In our example, all the details are necessary since it is designed this way. However, there could have been details that are not needed. Nevertheless, the server will have many other lines in the logs that are not important in this case. So, we have to skip those lines and only extract the ones needed.

3.2.2 Sample the Data

To write parsing rules, minimum data are carried from the logs from one of the servers. In our case, we have a python flask application where it logs to /data/logs directory. Imagine we have ssh access to the Linux server and read permissions on the directory and log files. The logs would be potentially zipped. Let us take a look at our directory and see what we have there:

```
[yusuf@flask-dev-01 ~]$ ls /data/logs/
flask-example-server.log-2020-01-06_01.log.gz
flask-example-server.log-2020-01-06_02.log.gz
flask-example-server.log
[yusuf@flask-dev-01 ~]$ less /data/logs/flask-example-server
  .log-2020-01-06_06.log.gz
[06/Jan/2020:13:37:39.708] ERROR log_example:hello_world:
  Could not load the previous items.
Traceback (most recent call last):
  File "/app/flask-logging-example/log_example/public/views.py",
  line 9, in hello_world
    raise Exception("Couldn't load the previous items.")
Exception: Could not load the previous items.
{}
request-log: 2020-01-06:13:37:39.709 dbdp.web01.dev.com
  212.129.72.174 GET ff62a4f3...
[06/Jan/2020:11:06:05.327] INFO log_example:recommended:
  Designing
[06/Jan/2020:11:06:05.327] WARNING log_example:recommended: Big
[06/Jan/2020:11:06:05.328] ERROR log_example:recommended: Data
```

One thing we can do is to copy/paste some of this data to another directory so that we can work on a parser that filters the access logs from the rest of the logs. After that, a parser could be implemented that parses the logs into the format desired.

3.2.3 Building the Shell Command

We took a look in the data, sampled it, and saved it to another log file. Now, let us get rid of everything else but the request logs as we want with the following command:

```
zgrep request-log /data/logs/flask-example-server
  .log-2020-01-06_06.log.gz | cut -c 14-
```

zgrep takes lines starting with request log, and *cut* removes the first 14 characters. These columns are parsed into a more available format like comma-separated values (CSV). So, the following *awk* command are executed where we print first 10 columns and the remainder as the last column. Since a unique cookie is assigned for each user, various phases of user activity can be seen in the website. By default, the logs would be sorted by the order of events, so sorting them manually will be unnecessary:

```
| awk '{s = ""; for (i=1; i<= NF; i++) s=i<12&&i>1 ?
  s",""$i:s" "$i; print s}'
```

The next step is finding ways to optimize the website. Looking at the log data, there are many reasons that make the website less attractive. The following analysis are considered to see if the user activity gets affected by different variables:

- Number of error counts per page view
- The ratio of activity between browser types

In the number of error counts, the status code is used to analyze error counts. In this command, the sixth and ninth columns are first extracted. Sixth column is for Http address, and the ninth column is for status code. Later, *sort* command is used to sort it by the page views, followed by the use of *uniq* command to get counts per address and status duo. We can now sort by page views; however, this would be unnecessary since users would obviously go to certain pages more often. A better way is probably getting an overall count and compare the ratios. For example, if we get more 500s for adding items to the basket, then we might try to improve this part:

```
| awk '{print $6","$7}' | uniq -c | sort -r
```

The next thing to do is to extract browser type along with the page view. This might be useful since there might be some difficulty in serving pages for certain browser types. This would be a bit more tricky when constructing the command since user agent parsing can be troublesome. In the first *awk* operation, the input are converted into a comma-separated duo, and any comma is removed from the second column. In the second one, we try to find a browser type by matching against well-known browsers. This is a best-effort script, and it only relatively solves the problem. The user agent parsing is not that easy, but it is believed to be good enough for this job:

```
| awk '{s = $6","; for (i=11; i<=NF; i++)
  {gsub(/,/,"",$i);  s=s" "$i;} print tolower(s)}'
| awk -F',' 'match($2,/firefox|windows|chrome|safari/)
  { print $1","substr($2, RSTART, RLENGTH)}' | \
uniq -c | sort -r
```

Now that we have collected some data at this point, we can write the output to a temporary file. Nevertheless, only the above commands in a subset of the data were executed when, ideally, we want to execute this against all data among all servers. In Section 3.2.4, we will see how it is done.

3.2.4 Executing the Shell Command

Two shell commands were created in the previous section, but the goal is to go even further than that. The commands work fine on their own, but executing all commands and saving the results locally are done. Once the results in the local machine are obtained, we can start merging them. The key part done in the previous section is the reduction of the size of the data by eliminating many logs, extracting data, and counting. We are able to download data locally because it got smaller. Let us see how we can execute our commands on all servers that we have our application running:

```
while IFS= read -r host; do
  ssh ${SERVER_USER}@${host} \
    zgrep request-log /data/logs/*.gz |
    cut -c 14- |
    awk '{s = $6","; for (i=11; i<=NF; i++)
      {gsub(/,/,"",$i); s=s " "$i;} \
      print tolower(s)}' |
    awk -F',' 'match($2,/firefox|windows|chrome|safari/)
      { print $1","substr($2, RSTART, RLENGTH)}' |
    uniq -c |
    gzip >~/log-data/${host}-log.csv.gz
done <hosts
```

A small script will be written that connects to each host, executes the commands previously discussed, and writes the result to a folder. It reads the hostnames from a file called hosts, but one can also feed the hostnames directly. Moreover, it zips the data because the goal is to decrease the amount of data. As a result, we get all the parsed and compiled data in one folder from all servers:

```
cat ~/log-data/*.gz | zcat | awk '{print $1","$2}' | \
awk -F ',' '{r[$2","$3]+=$1}END{for(i in r) print i","r[i]}'
```

Since all of the data is now locally available, there will be an attempt to merge all of the them into one file using *cat* and *awk* commands. The cat command outputs all of the files into one stream, and *zcat* uncompresses them. The first *awk* statement simply rewrites the values as comma-separated values. The second *awk* statement counts the sum of the unique path and browser fields.

3.2.5 Analyzing the Results

All the data needed are now merged and readily available to use. We have either the path, status code, and count or the path, browser type, and count. It is interesting to see if there are relatively higher ratios between the pages concerning status code and browser type and if there are more errors for certain paths or if some pages are less frequently visited with certain browser types:

```
# percentages by browser type or status code
| awk -F ',' '{s+=$3;r[$2]+=$3}END
  {for (i in r) printf "%s\t%s%%\n",i,100*r[i]/s}'
# percentages by path
| awk -F ',' '{s+=$3;r[$1]+=$3}END
  {for (i in r) printf "%s\t%s%%\n",i,100*r[i]/s}'
```

Let us find the percentage of status codes or browser types. In the first *awk* statement, we remove the path and only count either browser type or status code. In the second *awk* statement, we simply sum all and print out percentages by dividing the count by sum:

```
| awk -F ',' '{b[$2]+=$3; p[$1]+=$3; r[$1,$2]+=$3}END
  {for (i in b) { for (j in p) {
  printf "%s\t%s%%\n",i","j,100*r[j,i]/b[i] } } }'
| sort -n -k2
```

Now that percentages are obtained, we can search for outliers in the data. For example, the paths that have much higher error status code are compared with the paths that were called much less with a certain browser type. The *awk* statement sums the results by browser type, path type, and combination of the two. Later, it finds percentages for any given browser type and path. The results are sorted by percentages afterward. We can now quickly see if there is an immediately available outlier. In browser type example, it shows that we do not have hit to some paths at all for Firefox.

3.2.6 Reporting the Findings

In Section 3.2.5, we looked at the data in different dimensions and found outliers. The analysis part can become much more elaborate depending on what we want to accomplish. We only looked at outliers, but one can go even further and check relative differences between paths and browser type or status code. The next step is reporting the findings to the development team. There might be various reasons why we do not have any hit for a certain browser type such as issues with cascading style sheets (CSSs) that makes it impossible to click a link. For status code, the service backing the path might have difficulties in responding, therefore eventually ending up in 500s:

```
| awk '{print $1","$2}' \
| awk -F ',' 'BEGIN { print "<table><tr>
<th>Browser</th><th>Path</th><th>Percentage</th></tr>" }
{ print "<tr><td>" $1 "</td><td>" $2 "</td><td>" $3 "</td></tr>" } END
{ print "</table>" }' |
mail -s "$(echo -e "Browser Percentages\nContent-Type: text/html")" \
browser-analyst@myCompany.com
```

The next step is reporting the findings to a distribution group where interested parties can get notified. We carry on our script by another *awk* statement that reorganizes the data with columns. The next *awk* statement creates an HTML table with prepared data by looping over rows. The *mail* statement simply takes the output from the *awk* statement and sends an email to the distribution list.

3.2.7   Automating the Process

The process was gone over step by step, and a brief outlook on the pages that users visit was constructed. Manually doing this every day might take time and is prone to failure. Hence, the better way to handle it is through automation. If we can run our script every day at a scheduled time, then we could easily automate the process. What is more is that the script should be potentially put to source control so that it is available for everyone and it does not get lost. Hence, here are the next steps to be taken:

- Write an entry point script
- Put both entry point scripts and analysis scripts to source control
- Add a cron job to process data every day:

```
REPO_PATH=/home/yusufaytas/
REPO_NAME=shell-analysis
cd "${REPO_PATH}/${REPO_NAME}" || cd "${REPO_PATH}"
git pull || git clone
 git@my.company.git.io:my-team/${REPO_ NAME}.git
bin/run_browser_analysis
bin/run_status_analysis
...
```

Our entry point script is quite straightforward. First, we navigate to the directory of the script. Then, we check for updates in the repository. After that, we execute analysis commands one by one. Analysis scripts can also be ran in parallel, but there is no need to put more load on the web servers:

```
00 00 * * * /home/yusufaytas/shell-analysis/bin/entry_point
```

Since scripts have been added to the source control, the next thing is to write the *cron* job to execute these scripts. Our simple *cron* job looks like the one above. Analysis scripts will run at midnight, so the results will be emailed to the interested parties in the morning every day.

Some small adjustments to our scripts need to be done so that they only process data for the previous day. Otherwise, we will end up processing all data every day. Moreover, there might be other parsing rules to be added as well. We have used a manageable example; nevertheless, the parsing can be much more laborious than the way it is handled.

3.2.8   A Brief Review

We built our data processing platform with the help of common Unix commands. Taking a moment to find the similarity between the approach we took and the MapReduce framework is necessary. Essentially, a bunch of commands were mapped to run on the web servers, and the results were reduced in another server. The processing is distributed since each command runs in another web server. What is more is that the processes could be made parallel with a couple of adjustments. It looks like we have the local server as a coordinator and web servers as slaves. The point is that the problem was dealt with in somewhat similar way and still got the results.

The solution we came up and MapReduce framework share a common pattern: **divide et impera**. Divide and conquer is a common approach that is used in many areas of computer science. With regard to Big Data, it is an indispensable strategy since processing so much data in one place is impossible. Throughout this book, examples of this strategy in different environments and frameworks will be shown.

3.3   Processing Large Data with PostgreSQL

In Section 3.2, a simple reporting system has been developed that informs the interested parties about several aspects of the web platform. That might be enough for some businesses or organizations, but some might need a deeper grasp. To address this problem, a solution is needed that offers a common interface to execute analysis tasks on the data. There are a few numbers of such solutions for addressing this need; however, it is believed that PostgreSQL would cater to most of the requirements. Here, PostgreSQL is used as our data processing engine for the following reasons:

- Open source that means no need to pay for the software license
- Basic table partitioning
- Common table expressions
- Window functions for aggregating data
- Unstructured data handling
- Leverages multiple CPUs to answer queries faster
- Well known with a large community

In this section, we will work on the same data as in Section 3.2. It is assumed that the data is parsed. Also, some scripting will still be written to load data into PostgreSQL though there is no need to revisit all the steps since they were covered.

3.3.1   Data Modeling

Previously, the data we get from request logs were described. Here, the goal is to create a model for that data in PostgreSQL. We will leverage PostgreSQL partitions to efficiently execute queries on given ranges.

Yet, before doing so, let us quickly visit PostgreSQL partitions. PostgreSQL partitioning splits one large table into smaller physical pieces. When PostgreSQL receives a query to a partitioned table, it uses different execution plans to take full advantage of individual partitions. For example, when appropriate, PostgreSQL can scan the whole partition instead of random access. Partitions can increase the gains from indexing when a table grows large because it is easier to fit the partitioned index to memory than the whole index. Moreover, PostgreSQL partitions come in handy since their bulk operation like inserting or deleting a partition is possible, avoiding a massive lock on the table. Those operations are very common in large data sets. We may need to put new partitions for each day or hour or, on the flip side, may need to delete hourly or daily partitions for retention purposes. Last but not least, partitions can be moved to other media devices without affecting the rest:

```
CREATE TABLE dbdp.request_log (
    ts                    timestamp WITH TIME ZONE NOT NULL,
    host                  text,
    ip                    text,
    http_method           text,
    unique_cookie         text,
    path                  text,
    http_status_code      int,
    content_length        int,
    request_time          int,
    http_referrer         text,
    user_agent            text
) PARTITION BY RANGE (ts);
```

Let us take a look at the data model. In the table, we have ts as shorthand for timestamp as the partition column and the rest of the columns parsed from the request log. All of the columns are nullable as some of them might be missed due to numerous reasons. Note that it might be a good idea to check the counts occasionally to avoid incomplete data.

3.3.2 Copying Data

When the data were processed with Unix tooling, most of the job were done on the web server itself. Nevertheless, it is ideal to use PostgreSQL for heavy lifting in this case. Thus, we have to transfer all parsed data to the PostgreSQL server and copy the data to PostgreSQL:

```
TODAY=$(date '+%Y-%m-%d')
YESTERDAY=$(date -d "yesterday" '+%Y-%m-%d')
TWO_WEEKS_AGO_PARTITION=$(date -d "2 weeks ago" '+%Y_%m_%d')
YESTERDAY_PARTITION=$(date -d "yesterday" '+%Y_%m_%d')
TABLENAME='dbdp.request_log'
```

```
#Only keep last 2 weeks of data
#Add yesterday's partition
psql -h ${POSTGRESQL_SERVER} -U ${POSTGRESQL_USER} ${DATABASE} -c \
  'DROP TABLE IF EXISTS
  '"${TABLENAME}_partition_${TWO_WEEKS _AGO_PARTITION}"';
  CREATE TABLE IF NOT EXISTS
  '"${TABLENAME}_partition_${YESTERDAY_PARTITION}"'
  PARTITION OF dbdp.request_log
  FOR VALUES FROM ('"'"${YESTERDAY}"'"') TO ('"'"${TODAY}"'"');'

while IFS= read -r host; do
  ssh ${SERVER_USER}@${host} \
    zgrep request-log /data/logs/*.gz |
    cut -c 14- |
    awk '{s = ""; for (i=1; i<=NF; i++) \
    { gsub(/,/,"",$i); s= i>10 ? s" "$i : s$i","; } print s}' |
    psql -h ${POSTGRESQL_SERVER} -U ${POSTGRESQL_USER} ${DATABASE} \
      -c "SET datestyle = 'ISO,DMY'; \
    COPY ${TABLENAME} FROM STDIN DELIMITER ',';"
done <hosts
```

In the first part of the script above, some of the variables are initialized to be used later. In the second part of the script, the partition is dropped from two weeks ago for retention purposes, and a new partition is added for yesterday's data. In the loop, we go over each one of the hosts and copy the data from each host to the PostgreSQL server. Note that there is no need to perform the extra parsing that was done before for the user agent. Ideally, the user agent string is kept as it is since it might be valuable for different purposes. When connecting to PostgreSQL, we assumed either we have .pgpass file or our IP is whitelisted by PostgreSQL server for authentication purposes. The authentication part will not be covered here as it is out of the scope of this book.

```
SET max_parallel_workers_per_gather = 4;
SELECT unique_cookie, count(unique_cookie) AS view_count
FROM dbdp.request_log
WHERE ts >= current_date - 15 AND ts < current_date
GROUP BY unique_cookie;
```

The data was loaded into PostgreSQL. Now, we can run our SQL queries over request log data. One important aspect is to limit queries to the partitions. Otherwise, PostgreSQL has to account for data for partitions that are not needed. This would be a common matter on queries over Big Data. Careful writing of queries is advised, and, in most of the cases, these are limited to ranges and partitions. We do not want to put an undesirable load on resources due to heavy queries.

In our example query, the view count per user are calculated by using a unique cookie. We also set *max_parallel_workers_per_gather* to enable faster query execution time by paralleling the work among worker processes. And that is just a start.

Detailed queries can be written to understand user behavior. We can potentially connect a business intelligence tool for our table and bring other tables from the production databases to join with the log data. Nevertheless, a single PostgreSQL server might not be enough to do such complicated thing. Thus, a brief look at the multi-server setup might be considered.

3.3.3 Sharding in PostgreSQL

Although one PostgreSQL might be enough for many workloads, it might not be sufficient for jobs that require more data. We can still stick to PostgreSQL by adding a couple of nodes to distribute data over multiple servers. Luckily, PostgreSQL has a feature called foreign data wrappers that provides a mechanism to natively shard tables across multiple PostgreSQL servers. When we run a query on a node, the foreign data wrapper feature will transiently query other nodes and return the results as if they were coming from a table in the current database.

3.3.3.1 Setting up Foreign Data Wrapper

The foreign data wrapper comes with standard PostgreSQL distribution. The wrapper *postgres_fdwm* is an extension in the distribution and can be enabled through the following command. Note that it has to be run by database admin:

```
CREATE EXTENSION postgres_fdw;
```

Once the extension is enabled, then a server can be created. In the remote servers, we expect to have a database as *dbdp*, user as *dbdp*, and schema as *dbdp* setup. Two servers can be created as follows:

```
CREATE SERVER dbdp_server_one
    FOREIGN DATA WRAPPER postgres_fdw
    OPTIONS (host 'one.pg.mycompany.com', dbname 'dbdp');

CREATE SERVER dbdp_server_two
    FOREIGN DATA WRAPPER postgres_fdw
    OPTIONS (host 'two.pg.mycompany.com', dbname 'dbdp');
```

To enable connection from the destination server to remote servers, there is a need to map users from the destination server to the remote servers. The mapping can be done as follows:

```
CREATE USER MAPPING FOR CURRENT_USER
    SERVER dbdp_server_one
    OPTIONS (user 'dbdp', password '*****');

CREATE USER MAPPING FOR CURRENT_USER
    SERVER dbdp_server_two
    OPTIONS (user 'dbdp', password '*****');
```

An easy way to finish the mapping is by importing the desired schemas from remote servers to the local server. The importing can be done as follows:

```
IMPORT FOREIGN SCHEMA "dbdp" FROM SERVER dbdp_server_one INTO dbdp_one;
IMPORT FOREIGN SCHEMA "dbdp" FROM SERVER dbdp_server_two INTO dbdp_two;
```

Now, we are done with pairing databases. One can pair all databases together so that each one of them can execute the same queries without any problem. Nevertheless, a master database is selected that delegates the query execution to the other servers. It might be preferred to set up a stronger machine in terms of CPU/memory for the master server. Next, we can figure out how the sharding is done.

3.3.3.2 Sharding Data over Multiple Nodes

Partitioning is already used when implementing the copy operation for the request logs. As we have discussed earlier, it provides advantages over a traditional table with regard to large volumes of data. The next idea is to go even beyond partitioning on a single server and partition data over multiple servers. Distributed partitioning is called sharding since it involves scaling out horizontally. Sharding is required when the amount of data for the table is getting close to the capacity of a single server. Besides, we might need parallel processing on multiple servers when answering analytics or reporting queries. There is just one caveat: always filter queries by partitions. Otherwise, the queries will soon exhaust the database system:

```
CREATE FOREIGN TABLE dbdp.request_log_2020_01
    PARTITION OF dbdp.request_log
    FOR VALUES FROM ('2020-01-01') TO ('2020-02-01')
    SERVER dbdp_server_one;

CREATE FOREIGN TABLE dbdp.request_log_2020_02
    PARTITION OF dbdp.request_log
    FOR VALUES FROM ('2020-02-01') TO ('2020-03-01')
    SERVER dbdp_server_two;
```

Assume that we have created our respective tables *dbdp.request_log_2020_01* and *dbdp.request_log_2020_02* in remote servers with the same definition discussed earlier. Now, a link should be created between the remote servers and the local server by treating the table on a remote server as part of the partition with the above statements. PostgreSQL also supports sub-partitioning that means partitions can be made even smaller. Following our example, daily partitions on remote servers can be made as follows:

```
CREATE TABLE dbdp.request_log_2020_02_01
    PARTITION OF dbdp.request_log_2020_02
    FOR VALUES FROM ('2020-02-01') TO ('2020-02-02');
```

```
CREATE TABLE dbdp.request_log_2020_02_02
    PARTITION OF dbdp.request_log_2020_02
    FOR VALUES FROM ('2020-02-02') TO ('2020-02-03');
```

With all sharding and partition, we can have a distributed PostgreSQL cluster that can be scaled out horizontally. Nevertheless, it requires additional development and maintenance to support partitions and sharding. Plugins can be used for partition management, or partition creation can be automated through scheduler. However, it would require additional effort. When it gets close to be on par with the complexity of a Big Data system, it might be a good idea to consider more scalable options.

3.4   Cost of Big Data

Running Big Data systems can be operationally very costly. Before jumping into the bandwagon, it is better to evaluate other options and to start with the most basic solution and add on to this solution until it is no longer sufficient for the company's needs. As you can see from the examples, a very basic reporting can be obtained through Unix tools. When it was not enough for thorough analysis, we proceeded to PostgreSQL solution and scaled it out horizontally. When the management of the servers gets tricky enough or the desired performance is not achieved, the Big Data solutions should be considered.

The Big Data tools require a different set of expertise that might not be common for most of the development stack. It is hard to find and retain talent with Big Data skills. Thus, the use of Big Data systems also come with the problem of expertise. One can outsource the Big Data system deployment and maintenance, and it might be a wise option. Nonetheless, both talent and outsourcing would cost an additional expense. If the company can cope with a more manageable setup, it might be worth to go on that route before going through the rabbit hole.

Turning down the use of Big Data systems is not suggested. However, I just think evaluating the requirements and choosing more comfortable alternatives if possible should be taken into consideration. Of course, if a company gains a significant competitive advantage using more diverse and near real-time data, then it is more reasonable to invest in Big Data infrastructure.

4

Big Data Storage

After reading this chapter, you should be able to:

- Identify the differences between different Big Data storage patterns
- Set up on premise data storage
- Use cloud data warehousing solutions
- Come up with a hybrid data warehouse
- Explain pros and cons of different solutions

In Chapter 3, few ways to deal with data were addressed. When the business requires more than simple solutions, we have to be ready. Before the emergence of big data, businesses already used solutions for analytics and business intelligence. However, these solutions are extremely expensive. Fortunately, the vast amount of data and advancement in distributed computing boosted the big data storage solutions. We will go through types of data storage and then discuss various technologies for different setups.

4.1 Big Data Storage Patterns

Over the years, storage patterns have changed drastically with regard to data processing. With the pace of big data, many concepts are presently obtained that are very close to each other but with subtle differences. The big data platform should effectively use relevant patterns to make the best out of the data. In this section, we will explain storage patterns: data lakes, data warehouses, and data marts.

4.1.1 Data Lakes

The appearance of the data lake concept is quite new. Dixon (2010) first referred to the term as a large body of water in a natural state. The catch here is that data

Designing Big Data Platforms: How to Use, Deploy, and Maintain Big Data Systems, First Edition. Yusuf Aytas.
© 2021 John Wiley & Sons, Inc. Published 2021 by John Wiley & Sons, Inc.

lakes may contain raw data without any processing and need further processing to make it worthwhile. This is, in fact, one of the defining differences between a traditional data store vs a data lake. In the ingestion phase, data can be stored without a schema. Later, the schema can be defined when the data is ready to be read.

The data lake can contain structured, semi-structured, and unstructured data both from internal and external resources. We can drive results from combining all of these data types. With regard to semi-structured and unstructured data, the parsing or defining of schema may be delayed until data is read. Different stakeholders may extract data from the same data source in various ways. Nevertheless, this has a cost in terms of processing and knowledge transfer. Extracting data might become a burden if done more than once. Thus, it might be a good idea to do the computation once and save the structured data once the data is read.

Since the data lake may contain data sources outside of the company, it might not be a good area for analysts. The domain knowledge in data lakes can get close to nonexistent. One might need to elicit the metadata from the data itself. Nonetheless, it might be a good sandbox for data scientists and other interested parties. I believe it is beneficial to document the metadata information once the data gets investigated for future use.

The data lake can serve as a source for other big data storage such as data warehouses and data marts. Once the data gets cleansed, it can be moved out of the data lake to finer controlled storage.

4.1.2 Data Warehouses

The data warehouses have been widely adopted by the industry for business intelligence and analytics. The data warehouses like data lakes can hold large amounts of data. Nevertheless, traditional warehouses cannot be horizontally scaled as can be with data lakes since the underlying technology between the two is different. Moreover, data warehouses accept well-defined data structures and often with some documentation. Companies typically require structured data since they drive business decisions based on the data. All decisions might be in limbo if the data structure or metadata is not well defined.

Data warehouses are the systems where different sets of structured data can be combined. The aggregation of the data forms the information that the business needs. The derived information becomes the organizational memory for years back reporting for several aspects of the business. We can go even further and set up a single source of truths from aggregated data. The single source of truth is an important asset that provides the same information for every department in the organization. A single source of truth helps organizations in making decisions based on the same facts.

Data warehouses provide ad hoc querying methods over different interfaces. Data analysts and engineers, as well as executives, can discover business insights from different viewpoints. Occasionally, the querying would result in insight into the business. So, it would become a dashboard over the underlying data serving different stakeholders.

4.1.3 Data Marts

A data mart is a relatively simple warehouse specialized in an area of the business or organization. For example, the company might have a data mart for sharing information with the partners. It is driven by subjects such as sales, customers, and partners and meets the expectation of different stakeholders within the organization.

Once the data gets processed in the data warehouse, one can import the data into the relevant data mart. Having a more focused use, data marts can be fast and easy to use. The reporting on top of data marts is also a general use case for them. If the company wants to have dashboards for sales, then it would make sense to feed data into a sales data mart and build dashboards on top of it.

4.1.4 Comparison of Storage Patterns

The comparison of different patterns of data storage and processing is presented in Table 4.1.

Table 4.1 Comparison of big data storage patterns

Properties	Data lake	Data warehouse	Data mart
		Business intelligence	
	Big data analysis	Reporting	Fast lookups for
Usage	Machine learning	Data analytics	summary data
	Data insights	Data visualization	
Data source	Internal systems	Data lakes	Data warehouses
	External systems	Relational databases	
	Structured	Aggregated	
Data types	Unstructured	Transferred	Summarized
	Semi-structured		

4.2   On-Premise Storage Solutions

An on-premise big data platform refers to deploying the platform to the company's data center, where the organization has to invest in hardware, maintenance, and talent. Thus, an on-premise infrastructure comes with an upfront cost for implementation. Nevertheless, an on-premise platform can cost much less than a cloud-based solution over the long term. One critical aspect of using on-premise solutions is whether the company already uses a data center or not. Investing just for big data platforms in the data center might not be a wise option. On the other hand, if the company already uses a data center, then it might be an appealing solution, particularly if there is much old hardware since they could be a cheap potential candidate for storage.

Hadoop Distributed File System (HDFS) is the de facto storage system for on-premise big data platforms that is widely used because of the initial tailwind plus the ecosystem around it. Hadoop makes it easy to use commodity hardware by assuming the routine of hardware failure. It replicates data among three nodes by default to make it fault-tolerant. Hence, HDFS and Hadoop ecosystem, in general, would be the choice for on-premise big data platform. Deciding on using Hadoop is the easy part. However, there are many aspects to be considered such as deployment, capacity planning, monitoring, and hiring. In this section, each of these aspects will be addressed.

4.2.1   Choosing Hardware

When setting up a new Hadoop cluster, it is generally good to start small and get a sense of the workload for a project. With this strategy, the team can get experience in the Hadoop ecosystem without making a notable investment in the hardware. All this in mind, we can choose hardware based on the nature of Hadoop. A single node does not play a big role in the cluster; the resiliency of the system comes from the overall cluster. Thus, there is no need for expensive hardware that has redundancy and resiliency for the most part. On the other hand, Hadoop clusters consist of different types of nodes. It is better to choose suitable hardware for NameNodes, job trackers, DataNodes, and application nodes like Hive Metastore.

4.2.1.1   DataNodes

DataNodes are the backbone of the cluster that are used for storage and MapReduce jobs. These nodes are expendable. Moreover, redundancy is not needed for these nodes since Hadoop takes care of it on its own. Instead of having redundant array of inexpensive disks (RAIDs), JBOD (just a bunch of disks) is used to decrease the cost. HDFS will handle the potential disk failures, so there is no need to worry about it. Hadoop also performs the computation on these nodes taking advantage of data locality. The nodes should have enough computation power concerning the disk density. Another important aspect is the space availability. If there are not many cabinets, then it is more reasonable to use big machines rather than cheaper ones since you can fit them into the same space.

4.2.1.2 NameNodes

NameNodes are critical for the availability of the cluster and are the single point of failure. It has all the information about the file system and knows the list of the blocks and their location for any given file in HDFS. Thus, a NameNode knows how to construct the file from blocks. It stores all these information in memory, requiring a large amount of it. One can set up more than one NameNode to achieve high availability where the nodes share the storage. The storage for NameNodes does not have to be bigger than the memory size since FsImage and EditLog as mentioned in Chapter 3 are relative to the size of the file system, thus the memory. Moreover, we should also use Network File System (NFS) to share state between NameNodes where only one of them is the primary and the others are standby.

In this setup, the failure for the NameNode would be handled manually. Still, we can take a step forward with the failover for NameNodes and have an automatic failover by using ZooKeeper. Since ZooKeeper itself is lightweight, it is okay to collocate ZooKeeper nodes with NameNodes. Nevertheless, we should have at least three ZooKeeper nodes for good coordination, hence three NameNodes.

4.2.1.3 Resource Managers

YARN needs one or more resource manager nodes to track resources in cluster and schedule applications. Resource managers are also a single point of failure for the cluster in terms of computation. Unlike the NameNodes, resource managers do not need so much memory but computational power since they are constantly negotiating resources with applications and tracking resources among node managers that live in DataNodes. Just like the NameNodes, resource managers can also be configured to obtain active and standby nodes to have high availability. With the help of ZooKeeper, resource managers can be configured to make automatic failovers. Again, ZooKeeper nodes can be collocated with resource managers.

4.2.1.4 Network Equipment

Hadoop uses the network to transfer files from clients to the DataNodes as well as MapReduce jobs. Moreover, it uses the network for status updates from DataNodes to NameNodes and node managers to resource managers. Choosing networking equipment depends on how much the data locality and the rest of the equipment available can be leveraged. If there are no powerful machines, then it is not appropriate to invest in costly networking equipment since saturation cannot be done. Another important aspect is about applications that are using the cluster, e.g. Presto or an external Spark cluster. Since these applications cannot take advantage of locality, more data need to be transferred to those nodes, hence requiring more bandwidth.

4.2.2 Capacity Planning

Once the node hardware is decided, then we can focus on the cluster to carry the expected workload and storage. The complexity of planning comes from the unknowns about current and future requirements. Since the Hadoop cluster is used as a data lake, it has to support a diverse set of job types with a conflicting set of resource needs. We will discuss capacity planning in terms of the overall cluster and resource sharing among the organization.

4.2.2.1 Overall Cluster

In general, the goal is to find an answer to the following questions. The better answers we have to these questions, the more accurate we can get:

1. What is the volume of ingested data for daily, weekly, and monthly?
2. What is the rate of data accumulation? Does it change a lot?
3. How many tasks each node is supposed to execute?

Taking these questions into account, the cluster capacity planning requires investigations for ingestion, compression, and retention.

Ingestion   Calculating the ingestion rate for the cluster requires an understanding of the trends and preparing for the peaks. The increase of overall data volume can be predicted by looking at the trends. Nevertheless, the peak times might require special-case handling since there can be sudden increases in the volume. Therefore, an extra capacity to handle the peaks is a must.

Compression   Compression is an important player in the amount of storage space needed when planning capacity. Some file types can be compressed fairly well like server logs, while others would not. Nevertheless, the media types might not give much compression as they might be compressed beforehand. Thus, it is important to conduct some experiments with different file formats and decide on an expected compression ratio.

Retention   Retention defines how much data we would like to keep around. Apart from regulatory reasons, it is advised to save storage space if possible. It is also important to evaluate if there is a need at all for storing some data if they are already summarized. If the data is well summarized, then maybe it is best to store it in tapes rather than in the Hadoop cluster.

4.2.2.2 Resource Sharing

Hadoop is a general-purpose infrastructure for most organizations but makes sharing resources quite tricky. If one user group overuses the cluster, then the others might suffer. Hadoop solves this problem with the concept of queues that are allocated a fraction of the capacity of the cluster. All jobs submitted to a queue will have access to the capacity allocated to the queue. So, if the sales organization wants to process more data, then they have to pay for their capacity. Moreover, the queues can also be hierarchical, so one can assign resources among the departments, e.g. ad hoc or service-level agreement (SLA).

Sharing compute resources is one part of the problem, and the shared storage is the other one. Following the same principle, the queues can also define the amount of the storage as well since they are proportional to the compute resources. When planning the capacity, it is better to divide requirements and justify the need by department, since a big bill for infrastructure might be too big for one department.

4.2.2.3 Doing the Math

Considering all the factors discussed previously, the formula is given to calculate how much storage needed so that the capacity planning can be executed accordingly:

$$dS = \frac{c * r * a * (1 + g)^y}{(1 - i) * (1 - b)} \tag{4.1}$$

where dS is the data storage; c is compression ratio; a is an initial amount of data; r is replication factor, typically three for production systems; g is growth rate, a function of data ingestion vs retention; y is yearly growth rate; i is the amount of intermediate space for MapReduce, typically between 20% and 30%; and b is the buffer for free space in the case of spikes. If we have 400 TB to store initially with a compression ratio of 0.6, a replication factor of 3, a yearly growth rate of 10%, a buffer of 20%, and an intermediate space factor of 25%, we would require 1597.2 TB space in three years:

$$dN = \frac{c * r * a}{d * n * (1 - b) * (1 - i)} \tag{4.2}$$

When it comes to calculating number of DataNodes, a part of the formula above is used. Hadoop allows us to add/remove nodes whenever needed. In the above formula, dN is the number of DataNodes, c is the compression ratio, a is the amount to be stored, i is the intermediate space, b is the buffer, d is the disk size, and n is the number of data disk per node. If we have 400 TB, with compression ratio of 0.5, a replication factor of 3, a buffer of 20%, an intermediate space factor of 25%, 10 for disk per node, and 4 TB as disk size, we need 25 DataNodes.

4.2.3 Deploying Hadoop Cluster

Hadoop administration requires expertise in many areas. It requires an understanding of networking, operating systems, and management tools. Each one of these components will be discussed, and settings will be recommended. Later, we would also recommend a way to build a Hadoop cluster.

4.2.3.1 Networking

Hadoop learns the existence of a DataNode through heartbeats. In the first heartbeat, the NameNode will learn about the IP address and hostname of the DataNode. To send the data to the NameNode, when a DataNode starts up, it tries to learn its hostname and IP address. Hence, each node in the cluster has to perform a forward lookup for its hostname and a reverse lookup for its IP address. Unfortunately, the calls for the lookups are platform-specific, so these cannot be relied on. Some configuration updates will have to be done to make it work.

There are two network configurations we can rely on. The first one is using */etc/hosts* entries to define each node in the cluster. The second option is to use DNS resolution. Of the two options, using the DNS server is recommended since the management of all the hostname information is accurately centralized in one place. Adding/removing nodes from the cluster would be less error-prone. The hostnames should be set as a fully qualified domain name (FQDN) since it is required by Kerberos authentication or Transport Layer Security (TLS) encryption. Moreover, Hadoop uses DNS services quite often. To reduce the load on the DNS server, it might be a good idea to enable name service caching.

4.2.3.2 Operating System

Hadoop Users Hadoop allows running any MapReduce job. To decrease potential security problems, the use of separate users for HDFS and YARN daemons is proposed. A typical setup would have *HDFS* user for YARN daemons and *YARN* for YARN daemons both in the *hadoop* group.

Kernel Configurations The Linux moves memory pages to swap space when the page is not accessed for some time even if there is enough free memory available. Swappiness can be controlled with *vm.swappiness* that is between 0 and 100. The higher the value, the more aggressive the kernel for swapping pages. Nevertheless, swapping might cause timeouts when the disk is busy in other file operations. We recommend setting swappiness to 1, telling kernel not to swap pages often but still give a chance to do so.

To enable large-scale data processing technologies, both open file limits and the number of processes for the *HDFS* and other users should be increased. The limits can be increased by updating */etc/security/limits* for each user.

File System Linux maintains metadata for access time about the records in a file system. Nevertheless, this might create a problem since even read operations might cause a write in file system. This can be avoided by simply remounting the system with extra parameters to disable it.

4.2.3.3 Management Tools

Hadoop cluster management requires constant attention and maintenance. Some many components and nodes can fail at any time. Immediate fixes is not required, but they still need intervention in different ways. Thus, the Hadoop community came up with two different tools to make Hadoop cluster management less complex: Apache Ambari and Cloudera Manager. Both of these tools provide similar features to make Hadoop clusters easy to manage. Apache Ambari, being an Apache project, is fully open source. On the other hand, Cloudera Manager is open source for the base version and adds extra features with premium versions.

4.2.3.4 Hadoop Ecosystem

Hadoop ecosystem is the platform for different types of big data management, data processing, data access tools, and components. Since the Hadoop cluster has been deployed, additional tools such as Hive, Pig, and Spark can be used. All these tools help in different areas of big data management. To use these applications, we simply need to pass cluster configuration to the applications. For example, for a Spark application, we need to set the *master* as *YARN* and include required XML configuration files. How they are used will be described in the following chapters.

4.2.3.5 A Humble Deployment

In this section, a basic setup for a Hadoop deployment will be illustrated. In reality, the deployment would require the settings mentioned and some additional tuning. We recommend automating most of the node preparation tasks through either scripts or software provisioning and management tools like Ansible. Now, let us start our deployment, and we will have five machines in this. One of them would be hadoop-master, and others would be Hadoop nodes. Vagrant is used to set up the environment in our lab environment as follows:

```
Vagrant.configure("2") do |config|
  config.vm.define "hadoop-master" do |master|
    master.vm.box = "centos/8"
    master.vm.box_version = "1905.1"
    master.vm.network "private_network", ip: "192.168.7.1"
    master.vm.provision "shell", inline: "echo hello from hadoop-master"
  end
  (1..4).each do |i|
    config.vm.define "hadoop-node-#{i}" do |node|
      node.vm.box = "centos/8"
      node.vm.box_version = "1905.1"
```

```
      node.vm.network "private_network", ip: "192.168.7.#{i + 1}"
      node.vm.provision "shell", inline: "echo hello from hadoop-node-#{i}"
    end
  end
end
```

Once the machines are ready, we can ssh into machines with node names as follows: *vagrant ssh hadoop-master*. We first would like to prepare Hadoop users. To do so, the following commands on all machines in the cluster will be executed:

```
groupadd hadoop
useradd hdfs
useradd yarn
usermod -g hadoop hdfs
usermod -g hadoop yarn
```

Now that we have users setup, Hadoop can be installed. Let us prepare a common directory for both Hadoop users, set permissions, download Hadoop distribution, and finally install Java:

```
mkdir /home/hadoop
chgrp hadoop /home/hadoop/
chmod -R g+rwx /home/hadoop/
mkdir /home/hadoop
cd /home/hadoop
wget https://mirrors.whoishostingthis.com/apache/hadoop/common/
 hadoop-3.2.1/hadoop-3.2.1.tar.gz
tar -xzf hadoop-3.2.1.tar.gz
mv hadoop-3.1.2 hadoop

yum install java-1.8.0-openjdk
```

Hadoop nodes need to communicate to each other. We can do this either with DNS or updating *etc**hosts* file. In our private network setup, *etc**hosts* file will be used, and the following entries inserted for each host:

```
hadoop-master   192.168.7.1
hadoop-node-1   192.168.7.2
hadoop-node-2   192.168.7.3
hadoop-node-3   192.168.7.4
hadoop-node-4   192.168.7.5
```

We should also enable master access from YARN and HDFS users to DataNodes. To do so, we should generate ssh key pair and copy public key to authorized keys in the other nodes as follows:

```
ssh-keygen -b 4096
less /home/hadoop/.ssh/id_rsa.pub
#copy less output to each node in the cluster including master
```

```
#to the following file path
~/.ssh/authorized_keys
```

Next, we will prepare Hadoop environment variables for both *YARN* and *HDFS* users as follows. Note that Java path can differ between minor versions:

```
echo 'PATH=/home/hadoop/hadoop/bin:/home/hadoop/hadoop/sbin:
 $PATH' >> ~/.profile
echo 'export HADOOP_HOME=/home/hadoop/hadoop' >> ~/.bashrc
echo 'export PATH=${PATH}:${HADOOP_HOME}/bin:${HADOOP_HOME}
/sbin' >> ~/.bashrc
#java location will be probably different, set it to correct
 location
echo 'JAVA_HOME=/usr/lib/jvm/java-1.8.0-openjdk-1.8.0.242
.b08-0.el8_1.x86_64/jre/' >> ~/.bashrc
#replace JAVA_HOME in hadoop-env.sh
vi ${HADOOP_HOME}/etc/hadoop/hadoop-env.sh
```

We configure *$HADOOP_HOME/etc/hadoop/core-site.xml* in all machines as follows and set the NameNode location to hadoop-master on port 9000:

```
<configuration>
    <property>
        <name>fs.default.name</name>
        <value>hdfs://hadoop-master:9000</value>
    </property>
</configuration>
```

We configure *$HADOOP_HOME/etc/hadoop/hdfs-site.xml* in all machines as follows. The *dfs.replication* configures the replication count. We set it to 3 to have all the data duplicated on the three nodes. Note that we should not set a higher value than the number of DataNodes in the cluster:

```
<configuration>
    <property>
        <name>dfs.namenode.name.dir</name>
        <value>/home/hdfs/data/nameNode</value>
    </property>
    <property>
        <name>dfs.datanode.data.dir</name>
        <value>/home/hdfs/data/dataNode</value>
    </property>
    <property>
        <name>dfs.replication</name>
        <value>3</value>
    </property>
</configuration>
```

We configure *$HADOOP_HOME/etc/hadoop/mapred-site.xml* in all machines as follows. We set YARN as the framework for MapReduce:

```
<configuration>
    <property>
        <name>mapreduce.framework.name</name>
        <value>yarn</value>
    </property>
    <property>
        <name>yarn.app.mapreduce.am.env</name>
        <value>HADOOP_MAPRED_HOME=$HADOOP_HOME</value>
    </property>
    <property>
        <name>mapreduce.map.env</name>
        <value>HADOOP_MAPRED_HOME=$HADOOP_HOME</value>
    </property>
    <property>
        <name>mapreduce.reduce.env</name>
        <value>HADOOP_MAPRED_HOME=$HADOOP_HOME</value>
    </property>
</configuration>
```

We configure *$HADOOP_HOME/etc/hadoop/yarn-site.xml* in all machines as follows:

```
<configuration>
    <property>
        <name>yarn.acl.enable</name>
        <value>0</value>
    </property>

    <property>
        <name>yarn.resourcemanager.hostname</name>
        <value>192.168.7.1</value>
    </property>

    <property>
        <name>yarn.nodemanager.aux-services</name>
        <value>mapreduce_shuffle</value>
    </property>
</configuration>
```

We configure *$HADOOP_HOME/etc/hadoop/workers* in all machines as follows. We list all workers:

```
hadoop-node-1
hadoop-node-2
hadoop-node-3
hadoop-node-4
```

Let us start HDFS:

```
hdfs namenode -format
#start cluster, you should see workers starting
start-dfs.sh
#see information about cluster
hdfs dfsadmin -report
#simply create a directory and maybe put a file
hdfs dfs -mkdir -p /user/hadoop
hdfs dfs -put dbdp.txt /user/hadoop
```

Let us start YARN:

```
start-yarn.sh
#list nodes
yarn node -list
#list applications
yarn application -list
```

Voila! Hadoop installation is completed. Note that this is just a lab example. Normal installation would require automation software such as Ansible or Terraform management solution, and optimizations. Nevertheless, this gives a sense of accomplishment.

4.3 Cloud Storage Solutions

Cloud computing has emerged in recent years in different areas of the software industry. Cloud providers implement a platform-as-a-service (PaaS) solutions over the Internet in various domains of computing. Organizations can buy whatever they need from a cloud service provider in pay-as-you-go model. PaaS offerings include servers, storage, networking, middleware, development tools, business intelligence, database management, and more. Organizations can complete all software development lifecycle (building, testing, deploying, and managing) with the PaaS provider. PaaS can allow experimenting with anything quite fast without many costs. It also attracts businesses because it reduces the planning for provisioning for the infrastructure. Furthermore, cloud providers implement infrastructure-as-a-service (IaaS) solutions over the Internet. These allow organizations to establish their own data infrastructure without having to physically operate it on their own. With the IaaS model, companies have much more control over the software and hardware. Nevertheless, the company has to deal with all the details of the environment from security to provisioning.

With both PaaS and IaaS, organizations can leverage cloud providers to design and implement big data platforms. Organizations can build a Hadoop cluster using virtualized hardware or Hadoop-based services. Moreover, in big data space, cloud providers offer services such as object storage, data warehouses, data marts, apache big data tools, and archiving. Using either platforms or infrastructure, a big data platform can be designed and implemented end to end. In this section, some of the alternatives to cloud computing for big data will be discussed.

4.3.1 Object Storage

Object storage or object-based storage is a data storage technique that handles data as objects, instead of hierarchical methods used in file systems (Mesnier et al., 2003). With object storage techniques, each object contains the data, metadata about the object, and a globally unique identifier. The object storage technique swiftly became a service offered by cloud providers. Object storage services provide RESTful interfaces and client tools to manage data (Zheng et al., 2012) and often provides availability, durability, scalability, cost-effectiveness, ease of use, and performance. When data sets get extremely large, object storage services can be used to store big data. We can store either all the data in object storage or part of the data through Hadoop adapters. Object storage can come in handy because you do not have to deal with the deployment of new nodes for storage but also SLA guarantees around the storage itself.

How do we make use of object storage with regard to big data? There are a couple of alternatives to do so. A Hadoop cluster can be created on top of virtualized machines, and the data can be stored partly in Hadoop and partly in object storage. Hadoop has adapters for well-known object storage services. Using these adapters, tables can be created in the Hadoop cluster. This would be a great option when we have enough computation power but do not have enough space. With this approach, older data could be moved to object storage, and fresh data to the Hadoop cluster for ease of access. It would require additional automation to organize the swap, but it could be quite effective.

Another option is using object storage as the primary storage by using a preconfigured Hadoop cluster by the service provider. In this setup, the data lives in object storage and will be read when there needs to be computation. The output can be saved to the predefined location in object storage. This setup has the additional benefit of not maintaining a Hadoop cluster at all. Since the data lives completely in object storage, we can create a huge cluster on demand, process data, write results back, and destroy the cluster. Depending on the costs, this might be a really good option to process data quite fast.

The last option is to have object storage arbitrarily store data and read data on the fly to process it. This is somewhat similar to the second option; however, this is always available regardless of the setup. One can process unstructured data outside of the system. Further, this option also gives us the ability to integrate data with organization partners. Data can be potentially shared through object storage. In any event, we should be a bit more careful with this option since there might be no metadata at all.

4.3.2 Data Warehouses

A cloud data warehouse is a managed database service optimized for analytics processing. With the help of cloud data warehouses, we can query petabytes of structured and semi-structured data using standard SQL. Cloud data warehouses seamlessly integrate with their respective object storage. We can load data from object storages to data warehouses, process it, and write it either to the data warehouse itself or back to the object storage. With cloud warehouses, we can scale and pay for storage and compute separately. The amount of storage and compute for diverse workloads can be optimized interchangeably. Moreover, the size of our warehouse can be selected based on our performance requirements.

The cloud data warehouses have a different architecture when compared with each other. Some of them require a cluster to be provisioned, some hide all complexity from the end user, and some offer virtualization of data warehouses. In these different sets of architectures, one aspect is common. They all leverage a scale-out strategy to distribute query processing of data across multiple nodes. Let us quickly take a look of some of the common architectural patterns for cloud data warehouses and their applications.

4.3.2.1 Columnar Storage

Column-oriented storage has been widely adopted with regard to big data analytics due to its highly complex queries on a large set of data. It is no surprise cloud service providers implement their offerings on top of column-oriented storage. Column-oriented databases store each table column separately as opposed to traditional database systems where entire rows stored one after another as shown below. Values belonging to the same column stored contiguously and compressed (Abadi et al., 2009). Column-oriented storage makes it fast to read a subset of columns by having large disk pages. Nevertheless, column updates or scattered reads over a column become excessive due to disk seeking times. Typically, reading all columns becomes inconvenient since each column needs to be read independently:

```
== Row oriented
1:Mary,21
2:Paulina,43
3:John,45
4:Suzanne,11
5,GZ,15


== Column oriented
Mary:1,Paulina:2,John:3,Suzanne:4,GZ:5
21:1,43:2,45:3,11:4,15:5
```

One of the most important advantages of columnar storage is compression. The column value locality makes it easier to compress data well. There are compression algorithms that are developed for columnar storage, taking advantage of the similarity of adjacent data to compress. In order to improve compression, one can sort columns so that a greater compression rate can be achieved.

4.3.2.2 Provisioned Data Warehouses

In this type of warehouse, a cluster has to be provisioned with a defined number of nodes. Each cluster is composed of two main components: compute nodes and a leader node. The leader node orchestrates the compute nodes and responds to the client applications. The leader node sends the computational instructions to the compute nodes and allocates a part of the data to each compute node. Compute nodes have their dedicated CPU, memory, and disk storage and store data in partitions and execute queries. Each table data is automatically distributed across compute nodes to enable scale out of large data sets. This strategy also helps to reduce contention across processing cores. The data is distributed with the help of the distribution key that allows it to be colocated on individual slices as shown in Figure 4.1. Colocation reduces IO, CPU, and network contention and evades the redistribution of intermediate results of the query execution pipeline (Gupta et al., 2015).

Data is written both to the compute nodes and to object store so that data is more durable as object store would have higher durability guarantees. Hence, the loss of data might occur in the window of replication. Both object store and compute nodes are available for reading hence the media failures would be transparent for the clients.

4.3.2.3 Serverless Data Warehouses

In this type of warehouse, once the data is ready, there is nothing to be provisioned. The warehouse is auto-scaling; it computes resource requirements for each query

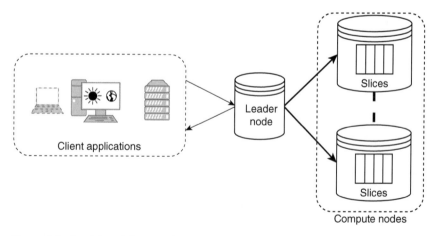

Figure 4.1 Provisioned data warehouse architecture.

dynamically. Once the query executions have completed, the execution engine reallocates resources to other warehouse operations. The queries run fast because the underlying engine has so much hardware available to devote to queries. The overall size of the underlying execution engine gives the ability to respond so quickly.

In order to achieve such impressive query performance, the execution engine dispatches queries and collects results across tens of thousands of machines in a matter of seconds. It uses a tree architecture to form a parallel distributed tree. The engine pushes down the query to the tree and then aggregates the results from the leaves. To do so, the execution engine turns SQL query into an execution tree. The leaves of the tree do the heavy lifting of reading the data from a distributed file system and doing the required computation as shown in Figure 4.2. Moreover, the branches of the tree do the aggregation. Between branches and leaves, a shuffling step is executed where the data is transferred (Melnik et al., 2010).

4.3.2.4 Virtual Data Warehouses

In this type of data warehouses, the architecture consists of three components: storage layer, compute layer, and services for managing data as shown in Figure 4.3. Separation of concerns makes it easy to scale different elements independently. This structure gives added versatility to choices for the underlying technology for each part. Moreover, implementing online upgrades becomes relatively more straightforward as opposed to node updates.

In this architecture, the system uses object storage to store the table data. Object store's performance might vary, but its usability and SLA guarantees make it an easy choice. The system uses an object store not only to store table data but also to store temporary data for intermediary results of the execution once local disk space is exhausted, as well as for large query results. Storing intermediary data in an object store also helps to avoid out of memory errors and disk problems.

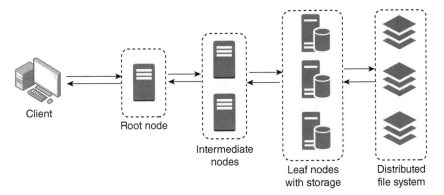

Figure 4.2 Tree data warehouse architecture.

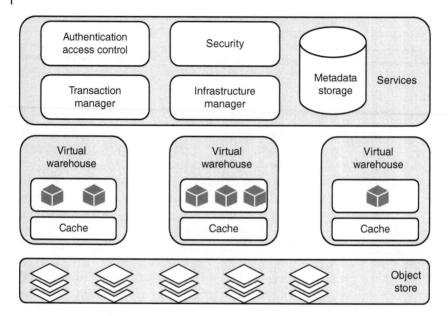

Figure 4.3 Virtual warehouse architecture.

For computation, the system uses the concept of virtual data warehouses. The virtual warehouse layer consists of clusters of computing resources. The clients never deal with the computing nodes. They only know the size of the cluster through an abstraction of cluster sizes named after T-shirt sizes. Virtual warehouses can be created, destroyed, or resized at any point, on-demand. Creating or destroying a virtual warehouse has no effect on the state of the system. The clients are encouraged to delete warehouses when there are no queries running.

The last component of the system is the services. The service layer provides access control, query optimization, and transaction management. The service layer is long-lived and shared across users of the system and is highly available since it is deployed to multiple nodes. The failure of service nodes does not cause outages or data loss. Nevertheless, it might create some problems with running queries (Dageville et al., 2016).

4.3.3 Archiving

Long-term big data retention might be required for several reasons from compliance, regulations, backups, and more. Cloud service providers implement achieving services to store seldom accessed data. These services optimize the underlying technology for infrequent data access and expect customers to access it occasionally. Instead of implementing an archiving solution in-house using tapes, it might be a cost-effective alternative to store large amounts of data in a cloud service provider. When using an archiving service, we can retrieve data in bulks within a couple of hours once it is saved to the archive. Customers are required to pay for the retrieval of data. A component in big data platforms can make use of archiving technologies to backup processed data.

4.4 Hybrid Storage Solutions

A hybrid big data storage is a combination of on-premise storage and cloud data storages integrated and can allow organizations to integrate data from the different systems and process them in a performant way. With a different set of technologies available on a cloud service provider, one can cost-effectively leverage different services. Moreover, some of the organizations have to keep some portions of their data in their data centers due to regulations, data privacy, industry standards, and more. In all these cases, a hybrid cloud infrastructure would be beneficial concerning scalability, adaptability, and flexibility of the big data platform. In this section, some of the hybrid solutions will be reviewed.

4.4.1 Making Use of Object Store

Previously, Hadoop adapters for object storage have been discussed. Using the adapters, we can expand on-premise clusters smoothly. That approach comes with a set of new techniques for both short-term and long-term storages. It also gives time to decide on what the next steps are if we have to take action quickly. Depending on the evaluation, we might either go fully in-house storage or continue investing in the hybrid infrastructure. Now, let us take a look at some of the hybrid usage patterns.

4.4.1.1 Additional Capacity
In case there is not enough disk capacity on our on-premise cluster, one quick way to solve the problem might be using object storage to extend it. Furthermore, we might use the extra storage for data that is not accessed frequently. Since the cluster has to stream the data from the service provider, it is better to restrict usage to the rarely accessed data. To do so, an extract, transform, and load (ETL) job that moves the data to the object store could be potentially created. Meanwhile, we can order additional hardware for storage purposes.

4.4.1.2 Batch Processing
Another alternative might be pumping the raw data directly to the object store where on-demand compute resources can be used to form a cluster, run series of

jobs, and write the result back to the object store. From there, maybe we could write the result back to the Hadoop cluster or other cloud services. This option gives us the versatility to run heavy ETLs on an ephemeral cluster and reserve computational resources of the in-house cluster to ad hoc analysis as well as lighter ETL processes. A disadvantage of this pattern might be transferring data. Ideally, data should be transferred as soon as it gets available. Otherwise, we need to wait for the completion of the transfer before we can launch the ephemeral cluster.

4.4.1.3 Hot Backup

Last but not the least, object store can be used as the backup method. If we are required SLA around the big data platform we are designing, then a backup process might be essential. At times, our data center might have a power outage, or a deployment might go south because of a configuration issue. As a result, we might not able to get any processing done for a while. Nonetheless, if the recent data can be mirrored in an object store, then the data could still be processed, and the key data sets can be delivered partially to the stakeholders. The business can still make decisions using this partial data. If we can even set SLA expectations for different data sets, then we could do an even better job for mirroring data.

4.4.2 Making Use of Data Warehouse

Different versions of cloud big data warehouses have been presented. Furthermore, there are a couple of advantages of using the data warehouses. Cloud-based data warehouses provide the elasticity to grow storage independently of computing resources. Hence, we can ingest the bulks of data without much of computing costs. Moreover, a cloud data warehouse can provide parallelism for queries on the same data set since computing and storage can be separate. We can easily run ETL jobs on the existing data set while end users can query it for ad hoc analysis without a performance penalty. What is more is that some of the data warehouse technologies cost only per usage. This might come in handy when the cluster does not have much to execute. Lastly, cloud warehouses have excellent durability and availability guarantees. If we could pump data in there, the key SLA data sets could still be delivered in most of the scenarios.

4.4.2.1 Primary Data Warehouse

Creating a scalable and high-performance data warehouse might be quite challenging. Adopting a cloud data warehouse immediately solves the aforementioned part of the problem. If we control the data lake, then we could manage to adapt to different data warehouse solutions. If some of the stakeholders have a strong preference for another cloud data warehouse, we can pump data from the data lake as part of periodic jobs. Nevertheless, it is better to have a primary data warehouse where historical summary data and recent data are kept. Having a single location to find out facts and insights about the organization would increase organizational knowledge.

The unpleasant part of embracing a cloud data warehouse is switching to another one. From data loading to the tools, many aspects might change significantly, especially the parts that require administration. The administration concepts might differ quite significantly since the architecture of the data warehouse might be completely different. Thus, it is better to evaluate a warehouse in several directions before choosing one. One important aspect is to take a look at the deliveries promised and delivered.

4.4.2.2 Shared Data Mart

As we have learned previously, data marts are specialized data warehouses that take a portion of data from a data warehouse targeted toward a specific audience. If the primary data warehouse is obtained in a cloud environment, then it is quite easy to spin off another one. We could potentially use the same data from an object store to power both data marts and primary data warehouse. Even if we need further processing on the data to make it ready for data marts, it would be still more convenient since integration between systems in the cloud would be smooth.

Another aspect would be using other technologies for data mart purposes. A data warehouse might be an expensive option. If we have technologies that support the same set of interface and cheaper than data warehouse option, we could use it. Again, it should be pretty straightforward transferring data between systems within the same cloud provider.

Just as importantly, we can share data marts with company partners. A cloud data mart would make it easy to share since the partner does not have to go through a firewall, VPN, etc. This option also makes the solution fall into a standard way since the cloud service provider would have standard documentation around the underlying technology.

4.4.3 Making Use of Archiving

Archiving the big data is quite hard as the data size grows exponentially. Using a cloud archiving method might be an excellent choice since it is cheaper. One important factor is to make archives easily accessible with some sort of order where data can be found effortlessly. Otherwise, we would have to stream more data than it is necessary. Since streaming data costs additional money, we should try to avoid it. If there are no regulations or business concerns about archiving the data in a cloud service provider, it is believed that cloud archiving is a strong option since we can hardly keep archiving as cheap as a cloud service provider.

5

Offline Big Data Processing

After reading this chapter, you should be able to:

- Explain boundaries of offline data processing
- Understand HDFS based offline data processing
- Understand Spark architecture and processing
- Understand the use of Flink and Presto for offline data processing

After visiting data storage techniques for Big Data, we are now ready to dive into data processing techniques. In this chapter, we will examine offline data processing technologies in depth.

5.1 Defining Offline Data Processing

Online processing occurs when applications driven by user input need to respond to the user promptly. On the other hand, offline processing is when there is no commitment to respond to the user. Offline Big Data processing shares the same basis. If there is no commitment to meeting some time boundary when processing, I call it offline Big Data processing. Note that I somewhat changed the traditional definition of offline. Here, offline processing refers to operations that take place without user engagement. The term "batch processing" was purposely avoided because operations in bulk for online systems can be performed. What's more, near real time Big Data might have to be processed in micro-batches. Nonetheless, we will focus on offline processing in this chapter.

Offline Big Data processing offer capabilities to transform, manage, or analyze data in bulk. A typical offline flow consists of steps to cleanse, transform, consolidate, and aggregate data. Once the data is ready for consumption, applications on top of the offline systems can use the data through interactive processing,

Designing Big Data Platforms: How to Use, Deploy, and Maintain Big Data Systems,
First Edition. Yusuf Aytas.
© 2021 John Wiley & Sons, Inc. Published 2021 by John Wiley & Sons, Inc.

reporting, and more. Bear in mind, interactive processing in this context might mean queries return data within 10 minutes. Offline Big Data processing might require several data sources to combine and enrich the data. When there is a stream of data arriving at the Big Data platform, we have to wait a certain amount of time to have enough data for offline processing. The typical waiting times are monthly, weekly, daily, or hourly. In short, the Big Data platform should offer a scheduling mechanism to process data periodically, as shown in Figure 5.1.

A typical offline Big Data processing can be used for the following scenarios where

- Data can be stale
- There are large data sets to process
- Business requires complex algorithms to run
- Latency can be in the order of hours or days
- Aggregation steps join multiple data sources
- Backup for online processing

For most of the offline processing, data staleness is rather expected. The staleness is also often required to pile up enough data to make good decisions. When the data freshness is essential, it is better to look for streaming Big Data solutions.

Offline Big Data processing might be run over all the data that the platform has. What's more, the aggregation steps might require multiple data sources. We can end up in high latencies for individual components when we run complex business logic or algorithms. Nonetheless, the important part is the capacity of the system rather than latency.

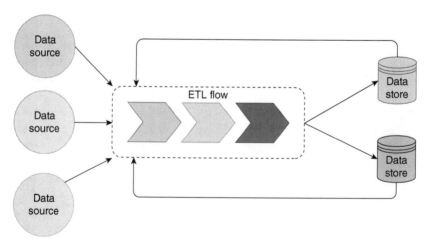

Figure 5.1 Offline Big Data processing overview.

Offline data processing might be used as a backup mechanism for online systems. Operationally things might go off the rails, and reprocessing of a series of events might become necessary. In such scenarios, offline processing might help to revise the data for online purposes.

Consequently, offline Big Data processing happens when the system does not have a time boundary for user input. The Big Data platform should accumulate enough data to process business logic periodically. Next, we will see some of the offline processing engines and leverage them in the Big Data platform.

5.2 MapReduce Technologies

The MapReduce framework has been a big step forward to process Big Data. Any modern Big Data platform might depend on the MapReduce framework. Although the MapReduce framework enables writing distributed applications easily, it requires understanding the programming model and Java. A steep learning curve makes it hard to develop Big Data applications for people who do not have a strong programming background. Moreover, the amount of code that needs to be written is relatively large. The applications might not have the best optimizations due to a lack of experience with the framework. For these reasons, both Yahoo and Facebook developed programming models over the MapReduce framework. We will now see how these models work.

5.2.1 Apache Pig

Apache Pig is a platform to analyze large sets of data represented as data flows. Pig provides a high-level language Pig Latin to write data analysis programs. Pig Latin offers functionalities for reading, writing, and processing data. Pig Latin scripts are converted to MapReduce tasks. Pig offers the following advantages over Hadoop.

- Pig makes it trivial to write complex tasks consisting of many data transformations representing inflow sequences. Pig data flow makes it easy to write, understand, and maintain.
- The way tasks are programmed makes it easy to optimize their execution to focus on the program rather than efficiency.
- Pig offers user-defined functions (UDFs) to extend the operations. UDFs make it simple to extend the core functionality.

Yahoo developed Apache Pig as a research project to create and execute MapReduce jobs on every dataset in 2006. Later in 2007, Yahoo open sourced Pig as an Apache incubator project. The first release of Pig came out in 2008. The last release of the Pig is dated 2017 as of the time of writing this book. Now, let us see how Pig works.

5.2.1.1 Pig Latin Overview

The Apache Pig takes a Pig Latin program as input, parses the program, optimizes the parsed program, and compiles it into one or more MapReduce jobs. Once the MapReduce jobs are ready, Pig coordinates those jobs on a given Hadoop cluster (Gates et al., 2009). Let us see an example program written in Pig Latin to find out top reviewed cities.

```
reviews = LOAD '/dbdp/reviews.csv' USING PigStorage(',')
  AS (city:chararray,userId:int,review:chararray);
reviewsByCity = GROUP reviews BY city;
mostReviewedCities = FILTER reviewsByCity BY COUNT(reviews)>1000000;
popularCities = FOREACH mostReviewedCities GENERATE group, COUNT(reviews);
STORE popularCities INTO '/dbdp/popularCities';
```

As you can see from the example program, Pig Latin is a data flow language where the programmer defines the transformations in each step. Each step uses variables from the previous steps to create the flow. A typical Pig Latin programs start with a *LOAD* command and end with *STORE* command to save the output of the execution. The example program also demonstrates the type of declaration where the programmer specifies the input schema for the data. A Pig Latin program goes through a series of phases before it becomes ready for execution shown in Figure 5.2.

5.2.1.2 Compilation To MapReduce

Pig translates a user program into a logical plan. Each statement in Pig Latin program becomes a step in the logical plan. Once the logical plan is ready, Pig translates it to a physical plan, as shown in Figure 5.3. As you can see, some of the logical groups became more than one step. In this context, rearranging means sorting by a

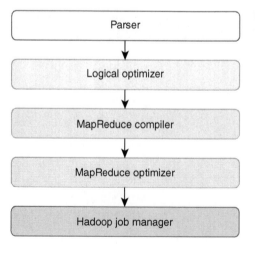

Figure 5.2 Pig compilation and execution steps.

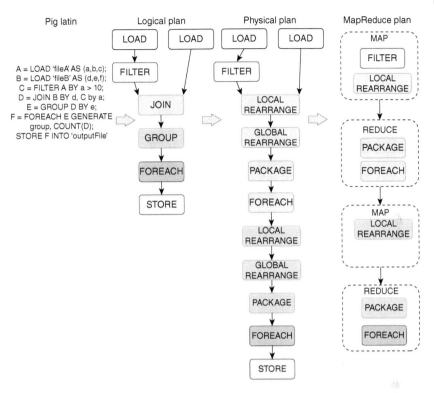

Figure 5.3 Pig Latin to MapReduce.

key locally/globally and distributing tuples to the same machine by the key. And, packaging groups the tuples by key. Later, Pig converts the physical plan into a series of MapReduce stages. The resulting MapReduce stages are more compact, and some of the operations like loading or storing do not exist since Hadoop takes care of them.

5.2.2 Apache Hive

Apache Hive, just like Apache Pig, is implemented over Hadoop to reduce the amount of effort for writing code. Hive uses a declarative language, Hive Query Language (HiveQL or HQL), which is similar to SQL, to ease the development of data processing (Thusoo et al., 2009). Out of the box, Hive supports many great things over plain MapReduce like data model, table partitioning, and driver support. With all this additional functionality, the Hive makes it straightforward for business analysts or data scientists to play with Big Data through ad hoc querying or scheduling. Let us now see the Hive deeper by drilling down to the Hive database.

5.2.2.1 Hive Database

We will first see how Hive Database models the data and later see hive query language.

Data Model Hive organizes data into tables, partitions, and buckets. Tables in Hive have a specified Hadoop Distributed File System (HDFS) directory. The user can specify the HDFS directory manually, but Hive will automatically assign one if it is not set. Hive stores the data for the table in a serialized format in this directory. Hive stores information about tables in the system catalog containing information about serialization format, HDFS directory, and more. It uses this information to query data from the table. Each table in the hive can have one or more partitions to create subdirectories for the table. Imagine we have a table stored in the HDFS directory */dbdp/example*. If the table is date partitioned, then we would have subdirectories like */dbdp/example/ds=2020-03-13*. The data in a partition can be further divided by buckets through a hash of a column.

Hive Query Language HiveQL supports most of the SQL capabilities such as joins, aggregates, unions, and subqueries. HiveQL supports data definition language (DDL) statements to create tables with different serialization formats to store the data as well as table partitioning and bucketing. Moreover, users can load external data from HDFS easily by defining an external table. HQL also supports UDFs and user-defined aggregation functions (UDAFs). Hence, it is very flexible in terms of abilities. An example DDL and query can be seen as follows:

```
CREATE TABLE IF NOT EXISTS dbdp.events_hourly (
    event_id    string,
    event_ts    timestamp,
    event_type  string,
    data        string
)
PARTITIONED BY (ds string, hh string)
STORED AS ORC;

ADD JAR /dbdp/dbdp.jar;
CREATE TEMPORARY FUNCTION is_bot AS 'dbdp.IsBot';
CREATE TEMPORARY FUNCTION to_hour_tring AS 'dbdp.ToHourString';

INSERT OVERWRITE TABLE dbdp.events PARTITION (ds, hh)
SELECT es.event_id,
    es.event_ts,
    es.event_type,
    es.data,
    es.ds,
    es.hh
```

```
FROM dldp.event_stream es
     INNER JOIN dbdp.filtered_events fe ON es.event_id = fe.event_id
WHERE NOT is_bot(es.user_agent, es.ip)
  AND ds = CURRENT_DATE
  AND hh = to_hour_string(HOUR(CURRENT_TIMESTAMP)-1);
```

In the example DDL, we create a table for hourly events streaming to our data pipelines. We partitioned the table by *ds*(date) and *hh*(hour). If the partition size is too big, one might also partition the table by *event_type*. In this example, we have used a UDF to eliminate bot traffic. We used a relative path from the machine, but it can be an HDFS path for the JAR. Furthermore, we have used another table to find out filtered events as we do not want to store everything. This example can be a typical hourly job that reads through a stream and persists the results once we cleanse the data through filtering.

5.2.2.2 Hive Architecture

Hive architecture consists of many components. On a high level, Hive has clients, drivers, and metastore. We will take a look at these components now, as shown in Figure 5.4.

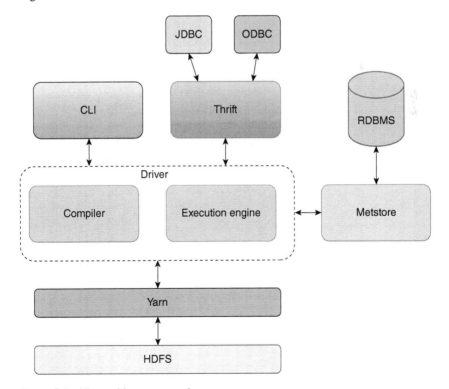

Figure 5.4 Hive architecture overview.

Hive Clients There are two ways to connect to the hive. First is through the command-line interface (CLI). The CLI allows the user to execute HQL files directly. The CLI also enables the user to use the tool interactively. The second option is through the Thrift server. Thrift enables cross-language communication with Hive. On top of the Thrift server, Hive has Java Database Connectivity (JDBC) and ODBC drivers to develop applications against the hive.

Hive Driver Hive driver consists of three main components: compiler, optimizer, and execution engine. The compiler transforms a given user query to a parsed tree representation. The compiler then does the semantic analysis where it checks types and type conversions. Once the query passes the semantic analysis, the compiler creates a logical plan which consists of a tree of logical operators. Hive then converts a logical plan into a query plan based on the execution engine then passes it for optimization.

Hive makes optimizations in two phases: logical query optimization and physical query optimization. The logical query optimization is based on relational algebra independent of the physical layer, while Hive employs several logical query optimizations like predicate pushdown, multi-way join, projection pruning, and more. The physical optimizations happen in the execution engine.

Initially, Hive only had MapReduce as an execution engine. Nevertheless, any Hive query produces a graph of MapReduce jobs. Employing a graph of MapReduce jobs might require writing to HDFS and can lead to many inefficiencies. New execution layers have been developed to address the shortcomings of the MapReduce framework. One of them is Apache Tez, which is a more flexible and powerful successor of the MapReduce framework. Tez has many improvements over MapReduce through multiple reduce stages, pipelining, in-memory writes, and more. Hive provides flexibility around the execution engine for better performance.

Metastore Metastore stores metadata for hive tables, partitions, and databases. Metastore provides critical information about data formats, extractors, and loaders. Metastore can be backed by a file or database. Nonetheless, a database would make a better choice. Furthermore, metastore can be either embedded or remotely connected through JDBC. We prefer remote metastore over embedded for administrative purposes. Metastore also provides a Thrift interface to connect.

5.3 Apache Spark

MapReduce has been widely adopted in Big Data processing. Nevertheless, MapReduce might be inefficient when the same set of data is used between different operations. Apache Spark addresses this problem while preserving the fault-tolerance and scalability of MapReduce. To achieve these properties, Spark introduced an abstraction called Resilient Distributed Datasets (RDDs). An RRD is read-only data set partitioned across multiple machines where a partition can be computed if lost (Zaharia et al., 2010). With this breakthrough, Spark has become a key part of Big Data processing and is a must-have for a modern Big Data platform. In this section, we will briefly touch on the ideas behind Spark.

5.3.1   What's Spark

Apache Spark is a unified analytics engine for Big Data processing. Spark provides a collection of libraries, APIs in multiple languages, and a diverse set of data science tools, graph processing, data analytics, and stream processing. Apache Spark can run on different platforms ranging from a laptop to clusters. It can scale to thousands of machines.

5.3.2   Spark Constructs and Components

On a high level, a Spark program has a driver that controls the flow of the application and has executors that execute the tasks assigned by the driver. The driver is responsible for maintaining application information and scheduling work for the executors. On the other hand, executors are responsible for executing assigned tasks from the driver. In doing so, Spark makes use of different constructs and components.

5.3.2.1   Resilient Distributed Datasets

An RDD is an immutable data structure that lives in one or more machines. The data in RDD does not exist in permanent storage. Instead, there is enough information to compute an RDD. Hence, RDDs can be reconstructed from a reliable store if one or more node fails. Moreover, an RDD can be materialized in both memory and disk (Zaharia et al., 2012). An RDD has the following metadata.

- **Dependencies**, the lineage graph, is a list of parent RDDs that was used in computing the RDD.
- **Computation function** is the function to calculate children's RDD from parent RDD.
- **Preferred locations** are information to compute partitions in terms of data locality.
- **The partitioner** specifies whether the RDD is hash/range partitioned.

Let us now see an example pyspark application where we would create an RDD out of a log file in HDFS. The data would be automatically partitioned, but we can repartition the data. Repartitioning data to a higher number of partitions might be important when distributing data to executors more evenly and potentially more executors to increase parallelism. In this example, we transform data two times by filtering. Later, we get the result back to the driver by counting.

```
from pyspark.sql import SparkSession
spark = SparkSession.builder.getOrCreate()
lines = sparksparkContext \
  .textFile ('hdfs://dbdp/dbdp-spark.log')
#transformation to filter logs by warning
warnings = lines.filter(lambda line: line.startswith('WARN'))
#transformation to filter warnings by connection exception
connection_warnings = lines \
  .filter(lambda line: 'ConnectionException' in line)
connection_warnings.count()
```

A typical spark application would load data from a data source, then transform the RDD into another one, as shown in Figure 5.5. A new RDD points back to its parent, which creates a lineage graph or directed acyclic graph (DAG). DAG instructs Spark about how to execute these transformations. After transformations are applied, the spark application finishes with an action. An action can be simply saving data to an external storage or collecting results back to the driver. Spark transformations are lazy. When we call a transformation method, it does not get executed immediately. Thus, the data is not loaded until an action is triggered. Transformations rather become a step in a plan of transformations, DAG. Having a plan comes in handy because Spark gets the ability to run the plan as efficiently as possible. There are two kinds of transformations: narrow and wide. A narrow transformation happens when each partition of parent RDD is used only by at most one partition of child RDD. Narrow transformations do not require shuffling or redistributing data between partitions. For instance: *map* or *filter* operations. A wide transformation happens when more than one partition of parent RDD is used by one partition of a child RDD. For instance: *groupByKey* or *join* operations. For wide transformations, Spark has to use shuffle operation since the data can be in several partitions of parent RDD. Narrow partitions allow operations to be pipelined in the same node. Moreover, narrow partitions help to recover efficiently since the transformation can happen in one node. In contrast,

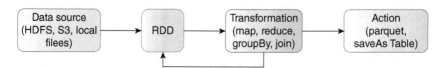

Figure 5.5   Spark RDD flow.

Narrow transformations   Wide transformations

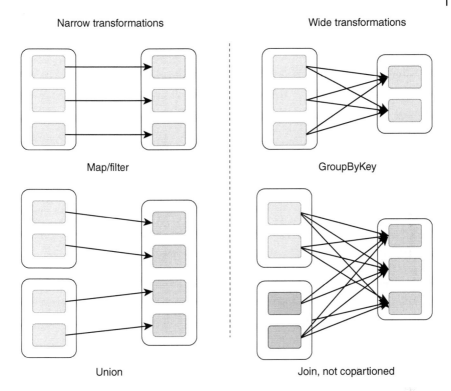

Figure 5.6   Narrow vs wide transformations.

a wide transformation node failure might require computing all of the data again.
Narrow and wide transformations are illustrated in Figure 5.6.

5.3.2.2   Distributed Shared Variables
Distributed shared variables are another element of low-level API in Spark. There
are two types of variables: broadcast variables and accumulators.

Broadcast Variables   The general way of using a variable throughout the cluster is
simply referencing the variable in the closures, e.g. *map/foreach*. Nevertheless, this
method might become inefficient once the variable size passes a certain threshold.
For example, a globally shared dictionary. This is where broadcast variables come
into play. Broadcast variables are a way to share immutable data across the Spark
cluster. Spark serializes broadcast variables and makes them available in each task.

```
airport_codes = {'IST': 'Istanbul Airport', 'DUB': 'Dublin Airport'}
broadcasted_airport_codes = spark.sparkContext.broadcast(airport_codes)
print(broadcasted_airport_codes.value)
```

Accumulators Accumulators provide propagating a value to the driver that is updated during transformations in the executors. We can use accumulators to get sum or count for an operation. Accumulator updates only happen in actions. Spark updates each value of an accumulator inside a task only once. We can use accumulators as follows:

```
def count_positive_review(review):
  if review.score > 0.75:
    positive_review_count.add(1)
positive_review_count = spark.sparkContext.accumulator(0)
reviews.foreach(lamdbda review: count_positive_review(review))
positive_review_count.value
```

5.3.2.3   Datasets and DataFrames

Spark has a concept of a dataset that represents a collection of distributed data. On a high level, datasets are very close to RDDs. But, they support a full relational engine. If we want to perform aggregation or SQL like filtering, we can rely on built-in aggregation operations instead of writing them. We can add to our previous example about log lines and try to determine the count for the different types of exceptions. Since dataset API is not available in Python, we will use scala instead.

```
spark.sparkContext.textFile("hdfs://dbdp/dbdp-spark.log").toDS()
  .select(col("value").alias("word"))
  .filter(col("word").like("%Exception"))
  .groupBy(col("word"))
  .count()
  .orderBy(col("count").desc)
  .show()
```

A DataFrame is a dataset with named columns. A DataFrame is synonymous with a table in a relational database. DataFrames are aware of their schema and provide relational operations that might allow some optimizations. DataFrames can be created using a system catalog or through API. Once we create DataFrame, we can then apply functions like *groupBy* or *where*. Just like RDD operations, DataFrames manipulations are also lazy. In DataFrames, we define manipulation operations, and Spark will determine how to execute these on the cluster (Armbrust et al., 2015). In the following example, we try to find the top visiting countries by using the visits table. As you can see, we are making use of SQL-like statements.

```
val visits = spark.table("dbdp.visits")
visits
  .where($"ds" === "2020-04-19")
  .select($"device_id", $"ip_country")
  .distinct()
```

```
.groupBy($"ip_country")
.count()
.sort($"count".desc)
.write.saveAsTable("dbdp.top_visiting_countries")
```

We can get the same results through Spark SQL as follows. Note that both options would compile to the same physical plan.

```
spark.sql("SELECT ip_country, COUNT(1) " +
  "FROM (SELECT DISTINCT device_id, ip_country " +
  "FROM yaytas.visits WHERE ds='2020-04-19')t " +
  "GROUP BY t.ip_country " +
  "ORDER BY 2")
```

The compiled physical plan is as follows:

```
== Physical Plan ==
*(4) Sort [count(1)#43L ASC NULLS FIRST], true, 0
+- Exchange rangepartitioning(count(1)#43L ASC NULLS FIRST, 10009)
   +- *(3) HashAggregate(keys=[ip_country#40], functions=[count(1)])
      +- Exchange hashpartitioning(ip_country#40, 10009)
         +- *(2) HashAggregate(keys=[ip_country#40],
            functions=[partial_count(1)])
            +- *(2) HashAggregate(keys=[device_id#41, ip_country#40],
               functions=[])
               +- Exchange hashpartitioning(device_id#41, ip_country#40,
                  10009)
                  +- *(1) HashAggregate(keys=[device_id#41,
                     ip_country#40], functions=[])
                     +- *(1) Project [device_id#41, ip_country#40]
                        +- *(1) FileScan orc dbdp.visits[ip_country#40,
                           device_id#41,ds#42]
```

5.3.2.4 Spark Libraries and Connectors

Being a unified execution engine, Spark supports different computational needs through its library ecosystem. We can use these libraries together. By default, Spark comes equipped with Spark streaming, Spark ML, and GraphFrames. Spark streaming enables Spark to process live data streams. The data can be ingested into Spark from many sources such as Kafka and Flume. Spark ML offers common machine learning algorithms such as classification, regression, and clustering. Lastly, GraphFrames implements parallel graph computation. GraphFrames extends DataFrame and introduces GraphFrame abstraction. Spark core enables these libraries to run on a cluster of machines and get the results fast. Please refer to Figure 5.7 for the relationship between the Spark core and its library ecosystem.

Spark is not a data store, and it supports a limited number of data stores. Instead, it offers a data source API to plugin read or write support for the underlying data

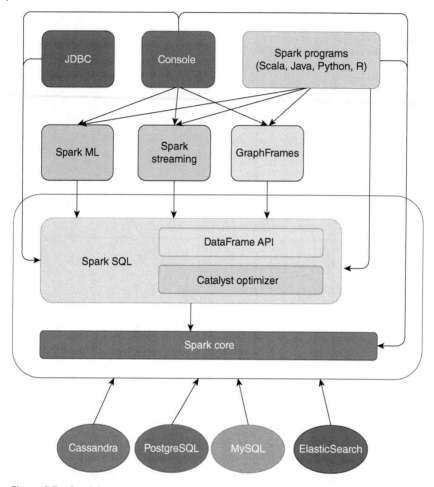

Figure 5.7 Spark layers.

source. By implementing data source API, one can represent the underlying technology as RDD or DataFrame. Thanks to the huge community of Spark, many connectors implement data source API.

5.3.3 Execution Plan

When a user submits a spark application, the code transposes into several phases and becomes actual execution. One of the parts of the execution is generating a logical plan, and the other one is the transformation to the physical plan, as shown in Figure 5.8.

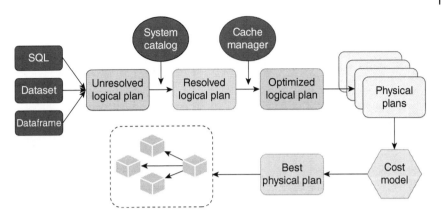

Figure 5.8 Spark execution plan.

5.3.3.1 The Logical Plan
The logical plan represents the transformations between child and parent RDDs. Spark first converts these transformations into an unresolved logical plan where it does not check metadata validity. Later, Spark uses the catalog to check the validity of the program. During this process, Spark tries to optimize the program as much as possible by constant folding, predicate pushdown, projection pruning, and null propagation.

5.3.3.2 The Physical Plan
After creating an optimized logical plan, Spark starts the physical planning process. In the physical planning process, Spark generates one or more physical plans and then compares them according to a cost model. Among physical plans, it chooses the most efficient one. The cost comparison might happen by looking at the attributes of a table. For example: when joining two tables, Spark might look at the table sizes to determine how to join them.

5.3.4 Spark Architecture

Spark relies on master/slave architecture where the master is the driver and executors are slaves. Spark driver controls the execution of a Spark application and maintains all state of the executors and tasks. It talks to the cluster manager yet another component to negotiate physical resources for running the spark application. Spark executors have only one job. Take the assigned task from the driver, execute it, report status, and return the result of the task. The cluster manager provides access to physical resources. Cluster manager also has a concept of driver and executors. Nevertheless, these concepts are rather tied to actual physical resources. When a Spark application starts, it first negotiates the

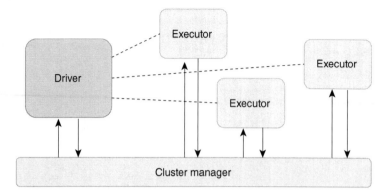

Figure 5.9   Spark high-level architecture.

resources through the cluster manager and then gets the necessary resources. The cluster manager then maintains the state information about actual machines. Currently, Spark has cluster managers for Kubernetes, Mesos, Yarn, and standalone clusters. Support for different environments is likely to grow. On a high level, Spark architecture is shown in Figure 5.9.

A spark application has three modes to run: local, client, and cluster. In local mode, Spark runs the entire application in one machine. It runs driver and executors in different threads. It supports the required parallelism through threads. Local mode is mostly useful for experiments and iterative development because it cannot scale much. On client mode, the Spark driver runs on the machine the Spark launches and runs the executors on the cluster. This mode might be useful when the driver needs to load additional data or configuration from the local machine. Lastly, a spark application can run on cluster mode. The only difference from client mode is that driver also runs on the cluster. The cluster mode is probably the most common way of running Spark applications.

5.3.4.1   Inside of Spark Application
A typical Spark application starts with initializing the SparkSession. SparkSession internally initializes SparkContext, which has the connection information to the spark cluster. Once the SparkSession is ready, we can execute some logical instructions. As we discussed before, a Spark application has transformation and actions. When an action is called in the application, Spark creates a job that consists of stages and tasks. Spark creates a job for each action. If an application has three actions, it would have three Spark jobs. Stages are a group of tasks that can be executed together to compute a series of transformations. Spark executes stages in many machines in parallel. Spark tries to pack as many tasks into one stage. Nevertheless, shuffle operations will cause new stages. A shuffle is the physical

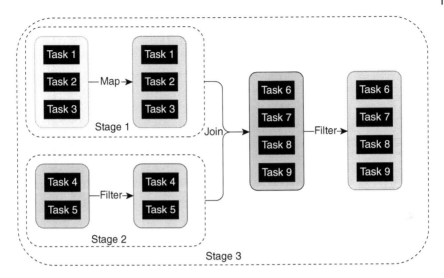

Figure 5.10 Spark stages.

repartitioning of the data. It requires data to be distributed among clusters differently. Tasks correspond to one more transformation and a block of data. The number of tasks depends on the number of partitions. If we have 19 partitions, then we would have 19 tasks. Figure 5.10 shows how Spark executes tasks in stages.

Spark computes jobs with the help of pipelines. Pipelining helps Spark to execute as much as it is possible before persisting results. Pipelining is one of the big advantages of Spark over MapReduce. When spark needs to shuffle data, it cannot pipeline the tasks any longer because it requires data to move between the nodes. In that case, it can save the intermediary results to memory or disk. Caching the data at the end of a pipeline helps when the next stage fails and requires the results from the previous stage.

5.3.4.2 Outside of Spark Application
Spark application starts with submitting a JAR file to the cluster manager. The request goes to the cluster manager, and then the cluster manager puts the driver onto one of the nodes available in the cluster. The driver starts running the Spark program by creating SparkSession. SparkSession talks to the cluster manager to start the executor programs. The number of executors, their memory is passed through *spark-submit*, e.g. *executor-memory*. The cluster manager starts executor processes in the cluster. At this point, the Spark job is ready to run. The driver and executors start communicating. The driver starts tasks on each executor, and executors respond to the driver. Once all tasks have finished, the Spark driver exists with either success or failure. The cluster manager releases the resources it uses.

Let us take a quick look at an example Spark job. In the following program, we specify a few options. Nevertheless, there are many options to control a Spark job. For a Spark job, *master*, *deploy-mode*, and *class* are essential to run the job.

```
./bin/spark-submit \
  --master ${MASTER_URL} \
  --deploy-mode cluster \
  --driver-memory ${DRIVER_MEMORY} \
  --executor-memory ${EXECUTOR_MEMORY} \
  --conf ${CONF_KEY}=${CONF_VALUE} \
  ${ADDITIONAL_CONFIG}
  --class ${MAIN_CLASS} \
  ${JAR_FILE} ${ARGS}
```

Figure 5.11 shows how a Spark job looks like from outside of the application. We have multiple nodes running driver and executor Java virtual machines (JVMs). Driver JVM has SparkContext, which keeps the information about the cluster state. Executors have tasks that are assigned to them.

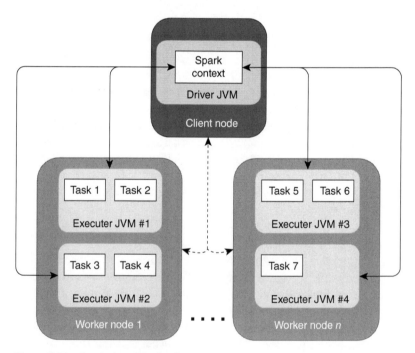

Figure 5.11 Spark cluster in detail.

5.4 Apache Flink

Apache Flink is an open-source stream and batch data processing engine. Flink's biggest feature is real-time data processing, where it offers high tolerance and low latency for distributed Big Data processing. Flink builds batch processing on top of stream processing. Flink considers batch processing a special case of stream processing where the input is bounded as opposed to unbounded. In this section, we will briefly touch on Flink and later discuss Flink in detail in Chapter 6.

Flink programs work on distributed collections by applying transformations such as filtering, mapping, updating state, joining, grouping, defining windows, and aggregating. We create Flink collections from data sources such as files and distributed storage. We complete the computation through sinks. Sinks can return the data through the console, writing to a file, and more. Flink can execute jobs in local JVM or a cluster of machines.

Flink supports batch and stream processing through DataSet and DataStream, respectively. Nevertheless, there is a plan to unite both APIs soon. Both DataSet and DataStream are immutable, distributed collections. We can create a collection through a data source and we derive new collections by transformations such as map and filter. Flink executes all transformations lazily. Each operation in Flink becomes part of the plan. Flink executes all operations when it executes.

Let us now see an example program where we would read a CSV file from HDFS. The file has visitor information, and we would like to get the visitor information grouped by age and country.

```java
public static void main(String[] args){
  //the file contains company,country,job,age
  final ExecutionEnvironment env = ExecutionEnvironment
    .getExecutionEnvironment();
  env.readTextFile("hdfs://dbdp/dbdp_visitors.csv")
      .map(new VisitorParser())
      .project(1, 3, 4)
      .groupBy(0, 1)
      .sum(2)
      .writeAsCsv("hdfs://dbdp/dbdp_visitors_grouped.csv",
        WriteMode.OVERWRITE);
}


public static class VisitorParser
    implements MapFunction<String, Tuple5<String, String,
      String, Integer, Integer>> {
  @Override
  public Tuple5 map(final String s){
```

```
    final String[] visitorAttributes = s.split(",");
    return Tuple5.of(visitorAttributes[1].trim(),
        visitorAttributes[2].trim(),
        visitorAttributes[3].trim(),
        Integer.parseInt(visitorAttributes[4]), 1);
  }
}
```

We can write a similar job by making use of SQL syntax as follows. We first create a temporary table and query it through SQL. Note that we also show the same statement through Table API in comments for the program. We can further have a Table API query on the Table object returned by a SQL query. Reversely, we can define a SQL query based on a Table API query by registering the resulting Table in the *TableEnvironment*.

```
//the file contains company,country,job,age
final ExecutionEnvironment env = ExecutionEnvironment
   .getExecutionEnvironment();
final BatchTableEnvironment tableEnv = BatchTableEnvironment
   .create(env);
// create an output Table
final Schema schema = new Schema()
    .field("company", DataTypes.STRING())
    .field("country", DataTypes.STRING())
    .field("job", DataTypes.STRING())
    .field("age", DataTypes.INT());
tableEnv
    .connect(new FileSystem().
      path("file:///dbdp/dbdp_visitors.csv"))
    .withFormat(new Csv().fieldDelimiter(','))
    .withSchema(schema)
    .createTemporaryTable("visitors");
tableEnv.toDataSet(
    tableEnv.sqlQuery("SELECT country, age, COUNT(age) "
        + "FROM visitors GROUP BY country, age"),
    new TypeHint<Tuple3<String, Integer, Long>>()
      {}.getTypeInfo())
    .writeAsCsv("file:///dbdp/dbdp_visitors_grouped.csv",
    WriteMode.OVERWRITE);
//same query can be written as follows
//tableEnv.from("visitors").groupBy("country, age")
//.select("country, age, company.count")
env.execute();
```

Flink is on the path of evolution, where the APIs would be unified in the later versions. In this book, I put examples that refer to Apache Flink version 1.10, but the API would be unified in the next major version. Nevertheless, the examples should not change drastically as the unified API would try to be backward compatible with the naming of the functions.

5.5   Presto

Presto is a distributed SQL query engine that supports various use cases such as ad hoc querying, reporting, and multi-hour ETL jobs. Presto emerged from the need for running interactive analytics queries. It can run against data sources from gigabytes to petabytes. Presto offers a connector API that integrates different data sources such as Hadoop, RDBMSs, NoSQL systems, and stream processing. It allows querying data on the data source it lives. A Presto query can execute on different data sources to get the relevant result to the caller. Enabling querying over different data sources decreases the burden on system admins as there are fewer systems to integrate. Moreover, Presto is an adaptive multi-tenant system that employs a vast amount of resources that scales to thousands of workers. It has different tuning settings that allow it to run a different set of workloads efficiently. It also offers optimizations like code generation to increase query performance (Sethi et al., 2019).

5.5.1   Presto Architecture

Presto has two types of nodes: coordinator and worker. A cluster has one coordinator and one or more workers. Coordinators accept the queries, parse them, plan and optimize the execution, and orchestrate the queries. Workers are responsible for query execution. Let us have a look at high-level presto architecture in Figure 5.12. A query comes into the Presto coordinator. The coordinator uses Metadata API to get relevant information from the external storage system. Later, the coordinator locates the data through Data Location API and schedules tasks. The workers fetch the data through Data Source API and run relevant logic on the data. Finally, the result is returned through the Data Sink API.

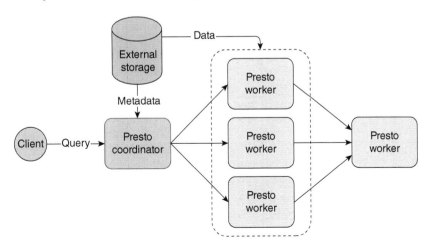

Figure 5.12   Presto architecture.

The coordinator accepts the query through an HTTP request and processes the request according to the queue policies. A queue represents the number of resources available for a particular query. Then, the coordinator calculates an optimized execution plan and shares the plan with workers. The coordinator starts the execution of tasks and enumerates them through data splits. Each split is a defined chunk of data that can be addressed in the external storage system. The coordinator assigns splits to the tasks.

Workers start tasks by fetching data for an assigned split or processing intermediate data produced by other workers. Data flow between tasks once the previous data gets processed. Presto pipelines the execution as much as it is possible. Moreover, Presto can return partial data for certain types of queries. Presto tries to keep the state and intermediate data in memory where it is possible to perform faster.

Presto is easily extensible. It provides a versatile plugin system. The plugins can be developed for various parts such as data types, access control, resource management, configuration, and event consuming. One of the most important parts of the plugin system is connectors. Connectors integrate Presto to external storage through the implementation of various API interfaces.

5.5.2 Presto System Design

Presto follows ANSI SQL with additional lambda functions. As Presto can connect to different data sources for a single query, various data sources should be called through their respective connector name. In the below query, Presto has two connectors: Hive and MySQL, referenced as hive and mysql, respectively. Upon receiving the query, Presto plans an execution, schedules tasks, and manages resources.

```
SELECT c.user_id, i.item_type, COUNT(c.user_id)
  AS count_by_item_type
FROM hive.dbdp.clicks c
LEFT JOIN mysql.dbdp.items i ON i.item_id = c.item_id
WHERE c.ds = '2020-05-06' AND c.created_at >
  CAST('2020-05-06 06:00:00' AS TIMESTAMP)
GROUP BY c.user_id, i.item_type
```

5.5.2.1 Execution Plan

Upon receiving the query, Presto parses it and generates a logical plan. A logical plan only contains information about logical or physical operations. For the query we shared earlier, Presto would generate a logical plan shown in Figure 5.13. As you can see, each node represents a logical or physical operation in the plan. Presto transforms the logical plan into a physical plan by applying transformations greedily until it finds the end.

After creating a physical plan, Presto starts optimization. The optimizer talks to connectors for additional information about the layout of data through the

Figure 5.13 Presto logical plan.

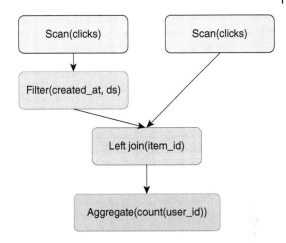

Data Layout API. With layout information, the optimizer can minimize the number of shuffles. The optimizer pushes down the filtering predicates to the connectors when it is appropriate. Pushing predicates can help reading only relevant information from the underlying data source. Furthermore, the optimizer tries to find out the parts that can be executed in parallel. These parts are called stages consisting of tasks. Stages for the logical plan for our example query is shown in Figure 5.14. The optimizer also tries to parallelize the processing in the node through multi-threading. Intra-node parallelization is more efficient since threads can share the memory, whereas workers cannot share it. In taking advantage of threads, the engine can execute a pipeline with additional local parallelism.

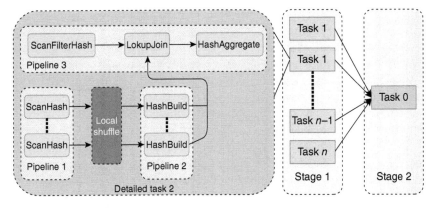

Figure 5.14 Presto stages.

5.5.2.2 Scheduling

The coordinator disseminates the execution plan to the workers in the form of tasks. A task is a base processing unit in the stages consisting of one or more pipelines. A pipeline is a sequence of operators where an operator is a well-defined function. The execution engine determines the scheduling of stages and tasks.

Presto executes stages in two different fashion: phased and all at once, configurable through presto config. In all at once case, Presto schedules all stages concurrently, which minimizes walk clock time. In phased scheduling, the Presto schedules stage in topological order. For example, when joining two tables, the table fetch operation happens one by one. Phased scheduling minimizes the amount of memory used at any time.

Presto has two types of stages: leaf and intermediate. Leaf stages are responsible for fetching data from data sources through connectors. The scheduler makes use of data layout to determine the best way to schedule tasks. The intermediate stages are stages between leaf and end-stage inclusive. For both stage types, Presto has to determine how many tasks it would use. Depending on the data layout or configuration, Presto can update the number of tasks dynamically.

Presto enumerates through data splits over tasks. The information for the split depends on the connector. Sometimes it might be a file path, and sometimes it can be connection information to hosts and table configuration. The coordinator assigns new splits to the tasks based on the queue size of the worker. The queues allow the system to adopt a performance difference between the workers.

5.5.2.3 Resource Management

Presto comes with a resource management system that allows multi-tenant deployments. A Presto cluster can handle many queries utilizing processing, memory, and IO resources. Presto optimizes for overall cluster utilization. Presto takes both CPU and memory into account when running queries.

Presto keeps track of CPU utilization both globally and locally. Locally, at the node level, Presto keeps track of a task resource usage by aggregate time spent on each data split. Presto schedules many tasks on each node in the cluster and used a cooperative multitasking model, where threads give back resources when the maximum time is reached for a given task. When a task releases resources, Presto needs to decide which task to run next. Presto uses a multi-level feedback queue to determine which task to run next. As tasks use more CPU time, they get higher in the queue. Furthermore, Presto gives higher priority to the queries with low resource consumption locally and globally. This allows inexpensive ad hoc queries to run quickly and return the results to the user.

Presto has the concept of memory pools. It has a system memory that is related to the system, e.g. implementation decisions. This memory type does not have direct relevance to the user query. On the other hand, user memory is available for users to consume. Presto applies global and node-level limits for queries. If a query exceeds the memory limit, either system or user memory, Presto will kill it. Per node and global memory limits are specific to the query itself. Presto allows queries to overcommit to the memory, i.e. allow queries to run even if there is not enough memory. Overcommitting is generally not an issue, as it is highly unlikely all queries would consume the max amount at a given snapshot in time. When there is not enough memory left, Presto stops tasks and writes the state to the disk. This is called spilling.

5.5.2.4 Fault Tolerance

Presto coordinators are a single point of failure. Manual intervention is needed if the coordinator goes down. If a node goes down during the query execution, the query fails altogether. Presto simply expects clients to retry the query in case of failure rather than handling failures on its own. Since standard checkpointing, partial recovery, and keeping track of transformations are computationally heavy, Presto simply does not implement these.

6

Stream Big Data Processing

After reading this chapter, you should be able to:

- Explain the need for stream processing
- Understand stream processing via message brokers
- Understand stream processing via stream processors

The first stop for Big Data processing was offline processing. In this chapter, we will examine stream data processing technologies in depth.

6.1 The Need for Stream Processing

As the number of devices connected to the network increases, so does the amount of streaming data. The streaming data comes from phones, tablets, and small computational devices. What is more, many systems produce data continuously. Some of the examples are the internet of things, social media, online transactions, and more. While data is coming in streams, the fast feedback loop is becoming increasingly crucial for applications and businesses.

Stream Big Data processing is a vital part of a modern Big Data platform. Stream processing can help a modern Big Data platform from many different perspectives. Streaming data gives the platform to cleanse, filter, categorize, and analyze the data while it is in motion. Thus, we don't have to store irrelevant and fruitless data to disk. With stream Big Data processing, we get a chance to respond to user interactions or events swiftly rather than waiting for more significant periods. Having fast loops of discovery and acting can introduce a competitive advantage to the businesses. Streaming solutions bring additional agility with added risk. We can change the processing pipeline and see the results very quickly. Nevertheless, it poses the threat of losing data or mistaking it.

Designing Big Data Platforms: How to Use, Deploy, and Maintain Big Data Systems,
First Edition. Yusuf Aytas.

We can employ a new set of solutions with stream Big Data processing. Some of these solutions are as follows:

- Fraud detection
- Anomaly detection
- Alerting
- Real-time monitoring
- Instant machine learning updates
- Ad hoc analysis of real-time data.

6.2 Defining Stream Data Processing

In Chapter 5, I defined offline Big Data processing as free of commitment to the user or events with respect to time. On the other hand, stream Big Data processing has to respond promptly. The delay between requests and responses can differ from seconds to minutes. Nonetheless, there is still a commitment to respond within a specified amount of time. A stream never ends. Hence, there is a commitment to responding to continuous data when processing streaming Big Data.

Data streams are unbounded. Yet, most of the analytical functions like sum or average require bounded sets. Stream processing has the concept of windowing to get bounded sets of data. Windows enable computations where it would not be possible since there is no end to a stream. For example, we can't find the average page views per visitor to a website since there will be more visitors and more page views every second. The solution is splitting data into defined chunks to reason about it in a bounded fashion called windowing. There are several types of windows, but the most notable ones are time and count windows, where we divide data into finite chunks by time or count, respectively. In addition to windowing, some stream processing frameworks support watermarking helps a stream processing engine to manage the arrivals of late events. Fundamentally, a watermark defines a waiting period for an engine to consider late events. If an event arrives within a watermark, the engine recomputes a query. If the event arrives later than the watermark, the engine drops the event. When dealing with the average page views problem, a better approach would be finding average page views per visitor in the last five minutes, as shown in Figure 6.1. We can also use a count window to

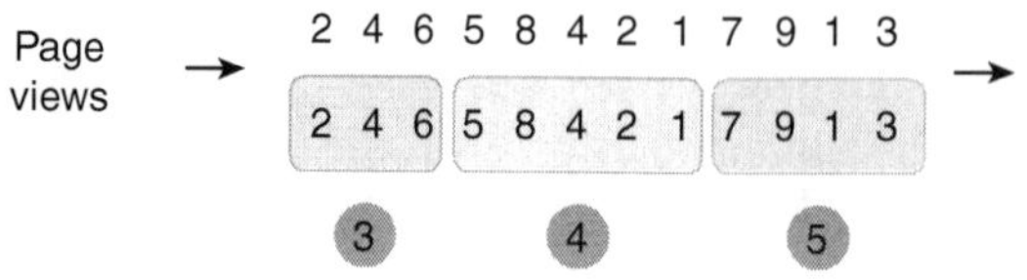

Page views → 2 4 6 5 8 4 2 1 7 9 1 3

2 4 6 | 5 8 4 2 1 | 7 9 1 3

③ ④ ⑤

Figure 6.1 Average page views by five minutes intervals.

calculate averages after receiving *x* number of elements. Many stream processing solutions use the windowing approach in different ways.

Several Big Data stream processing solutions emerged in recent years. Some of the solutions evolved from a messaging solution into a stream processing solution. On the other hand, some of the solutions directly tackle stream processing. In both cases, stream processing solutions consist of libraries and runtime engines. They allow developers to program against streaming data without worrying about all the implementation details of the data stream.

One breed of stream processing solutions comes from message brokers. Message-oriented middleware has long existed. The middleware allows components of a system to run independently by distributing messages from senders to receivers. The concepts might slightly differ between technologies, but they pretty much implement the same pattern, publish/subscribe. As messages flow through the message brokers, they create a stream of data. Thus, message brokers become a natural fit for Big Data stream processing with some additional effort. In Figure 6.2, I have shown a simple message broker. We can implement a streaming solution on top of the messages between the services. For example, we can have a fraud service that queries streaming data and finds out fraudulent transactions every five minutes. Then, we can cancel orders and ban fraudulent users.

The second breed of stream processors comes from the need for a stream processing engine. Message broker solutions provide the ability to process streaming data through messages. Nevertheless, they do not address data sources other than messages themselves. For this reason, we might need a general-purpose stream processing engine where it can accept different sets of data sources. What is more, the stream processing engines can still accept messages as data streams from message-oriented middleware through their connectors. For a broader spectrum of problems, a dedicated stream processing engine becomes a natural solution.

In the rest of the chapter, we will first discuss stream processing through message brokers. Later, we will go through stream processing engines.

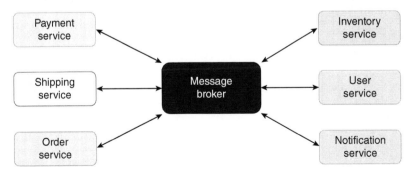

Figure 6.2 A message broker.

6.3 Streams via Message Brokers

As we mentioned earlier, message brokers have long existed. Message brokers enable applications to communicate with each other without direct dependency. Message brokers can validate, store, and route messages from one publisher to many subscribers. Message brokers can also support message queues, but we are not interested in that aspect in this book. We will focus on the publish/subscribe mechanism for messages or events. In Big Data stream processing, messages and events become a stream of data. Since we receive messages continuously, it is natural to implement an engine that enables stream processing. By taking advantage of the existing messaging infrastructure, several Big Data streaming solutions are implemented. In this section, we will cover popular message solutions implemented over Kafka, Pulsar and briefly mention brokers that adopt AMQP (advanced message queuing protocol).

6.3.1 Apache Kafka

Apache Kafka is a distributed streaming platform where it implements a publish/subscribe pattern for a stream of records. Kafka enables real-time data pipelines where applications can transform data between streams of data. Kafka accepts and distributes data over topics. Publishers publish to topics, and subscribers subscribe to topics. Kafka has a core abstraction on record streams and topics. A topic is a category where a record is published. A topic has zero to many subscribers. Kafka keeps a partitioned log structure for each topic shown in Figure 6.3. Partitioned logging helps to scale horizontally in terms of storage and parallelism. Each partition consists of an ordered and immutable sequence of records. Kafka appends new logs to this structured commit log. Each record gets a sequence ID, which helps to identify a record within a partition. Kafka persists records regardless of the subscribers with a retention period. Kafka only

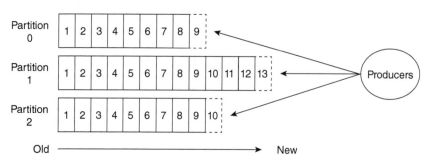

Figure 6.3 Kafka topic.

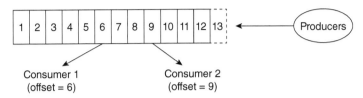

Figure 6.4 Kafka offset.

maintains offset, sequence id, for any given subscriber shown in Figure 6.4. Thus, keeping more logs does not affect performance. Kafka distributes partitions over several machines where each gets partitions for different topics. Kafka replicates a partition over a configurable amount of machines for fault tolerance. Each partition has a leader and followers. The leader partition accepts all read and write requests for the partition and then asynchronously sends logs to the followers. In the case of a leader failure, Kafka will elect a new leader among followers. Since a machine hosts many partitions, each node in the Kafka cluster might be a leader for some partitions. Having leaders in different devices help to distribute the load over multiple machines. Moreover, Kafka can replicate messages across data centers.

Kafka producers publish records on topics of their own choice. Kafka producer is responsible for which partition a record should go. It can choose a strategy for choosing a partition like a round-robin. On the flip side, consumers have a consumer group label, and a consumer gets a record in a consumer group. Consumer instances can live on the same server or another one. Kafka allows load balancing within a consumer group. Kafka divides the partitions between the nodes in the consumer group to provide a fair share of records to each consumer, as shown in Figure 6.5. Kafka maintains the membership of consumer groups. If a node comes or leaves, it rebalances consumption among the consumers. Note that nodes in consumer groups are free to subscribe to topics of their choice. Nevertheless, this might be tricky while trying to establish load balancing between consumers. Thus, it is easier to reason when consumers within a consumer group subscribe to the same topics. Kafka does not provide a total order of messages. It guarantees order within the partition. Still, one can achieve total order through one partition (Apache Kafka, 2020).

Now that we covered an overview of Kafka, let us see Big Data stream processing over Kafka's infrastructure.

6.3.1.1 Apache Samza

Apache Samza is a stream processing engine originally built on top of Kafka. Over the years, Samza added support for processing any data stream which has a changelog. Samza supports stateful processing. It can quickly recover from

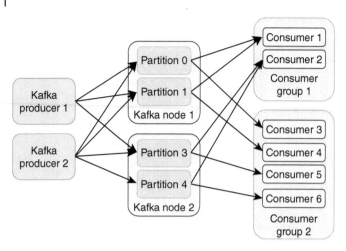

Figure 6.5   Kafka producer/consumer.

failures and resume processing. Samza can reprocess data streams quickly in case of bugs and problems in the data set. Samza utilizes a local state with a replicated change log mechanism to scale massive state sizes. Samza also allows processing a finite dataset without changes to application code, unlike lambda architecture (Noghabi et al., 2017). In lambda architecture, one has to have two pipelines where one pipeline is for stream processing and the other pipeline for batch processing. Nevertheless, this causes mismatches between stream and batch processing.

Samza processes data in streams, e.g. records in a Kafka topic. A Samza application takes a stream as an input, applies transformations and outputs as another stream, or writes it to permanent storage. When processing streams, Samza can do both stateful and stateless processing. Stateless processing means computing and handing over the result, e.g. filtering. On the other hand, stateful processing means aggregating data at intervals, e.g. the unique users in the last five minutes. Let us now build a sample Samza application where we filter out traffic from some IP addresses and then count page views per user in the last five minutes as follows:

```
pageViewStream
    .join(ipLookupTable, new JoinFn())
    .filter(pageViewWithLookup -> !pageViewWithLookup.
        getValue().getValue())
    .map(pageViewWithLookup ->
        KV.of(pageViewWithLookup.getKey(), pageViewWithLookup.
            getValue().getKey()))
    .partitionBy(pageViewKV -> pageViewKV.getValue().userId,
        pageViewKV -> pageViewKV.getValue(),
        pageViewSerde, "userId")
```

```
.window(Windows.keyedTumblingWindow(
    kv -> kv.getKey(), Duration.ofMinutes(5),
    () -> 0, (m, prevCount) -> prevCount + 1,
    new StringSerde(), new IntegerSerde()), "count")
.map(windowPane -> KV.of(windowPane.getKey().getKey(),
    windowPane.getMessage())))
.sendTo(outputStream);
```

We first join the stream to get a key-value of request-ID to page view with key-value of IP address to Boolean, where boolean indicates whether IP is acceptable. We then filter the page view if it does not have an accepted IP and map it back to the key-value of request-ID to page view. Later, we use a tumbling window to divide a stream into a set of contiguous five minutes intervals. Within intervals, we compute page view count for a given user id. Finally, we save the generated stream to the output stream.

Samza application is called a job where it consumes and processes a set of input streams. A job consists of smaller units called a task to balance out large amounts of data. Tasks consume a partition from the streams of data. Samza groups tasks into executable units called containers. A container is a JVM process that executes a set of tasks for a given job. Samza containers are responsible for starting, executing, and shutting down one or more tasks.

Processing Pipeline   A Samza job receives one or more streams, and various operations, e.g. filter, map, join, are applied. Samza represents jobs as a directed graph of operators as vertices and connecting streams as edges. A stream in Samza is continuous messages in the form of key, value pairs. Samza divides a stream into multiple partitions. An operation is a transformation of one or more streams.

Internally, as we described earlier, a job is divided into multiple tasks that receive input stream supplied in partitions. Each task executes the same logic but on a different partition, as shown in Figure 6.6. Each task executes the entire graph of operators. Tasks process each message through all operators until there is no operator left or output stream is reached. Intermediate results for streams stay local. Locality minimizes strain on the network. Nevertheless, Samza has to transfer between containers when it has to repartition the data. When partitioning happens, Samza employs replayable and fault-tolerant communication over Kafka.

Samza supports both in-memory and on-disk stores for local storage. The in-memory approach is fastest but is limited by the memory amount. Samza uses RocksDB for disk-based state management. Inside RocksDB, Samza keeps the state partitioned with the same partitioning function it uses for tasks. Moreover, Samza makes use of LRU caching to store the most frequently accessed items.

Storing state requires handling fault tolerance and easy recovery. Samza handles state management like a traditional database, where it stores a changelog

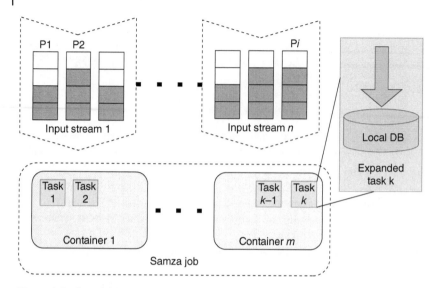

Figure 6.6 Samza job structure.

after each message. The changelog is an append-only structure where it only stores incremental changes to the state. Samza uses Kafka to store changelog in a fault-tolerant manner. Using Kafka means ease of replaying as well since you can restream data from a specific offset. Samza sends a changelog to Kafka in batches to avoid any degradation of performance during message computation.

Samza Architecture Samza has three layers: streaming, execution, and processing, as shown in Figure 6.7. The streaming layer is responsible for durable and scalable streams across multiple machines. The execution layer is responsible for scheduling and coordination of tasks across containers. Finally, the processing layer is responsible for the transformations of the stream. Samza uses Kafka for the streaming, YARN for execution, and Samza API for processing. Nonetheless, the implementation of streaming and execution are pluggable. We can use Apache Kafka, AWS Kinesis, and Azure EventHub for the streaming layer. On the flip side, we can use Apache Mesos or YARN for execution.

Samza provides fault tolerance through restarting containers. When a failure occurs, Samza recovers from it by replaying messages by resetting the offset to Kafka's latest successfully processed message. Nevertheless, replaying comes with a disruption in real-time computation since we might have to replay an entire day depending on the failure and operation. To address such challenges, Samza has a host affinity mechanism. The host affinity mechanism leverages the state stored in RocksDB rather than restarting the tasks.

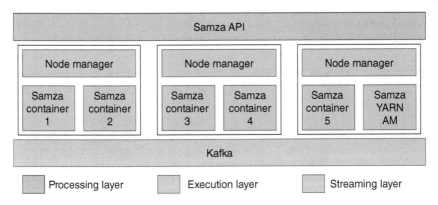

Figure 6.7 Samza architecture.

Nevertheless, this is an improvement over replaying. If the state is not recoverable through RocksDB, then this mechanism would not be sufficient. Hence, Samza has to replay messages instead of recovering the state. Samza supports an at-least-once processing guarantee because failure might end up reprocessing the same message again.

Since Samza leverages Kafka for streaming, a failure in downstream jobs would not halt the whole system. Upstream jobs can still execute without any issue since the results are flowing through Kafka. The downstream job can catch up at its own pace without causing any problem to the upstream job. Samza stores intermediate messages even if a job is unavailable. This gives a very flexible mechanism to deal with temporary hiccups.

Samza combines stream processing with batch processing by treating batch processing as a special case of stream processing with an end of stream token. Instead of receiving a stream from Kafka, it can consume the messages from HDFS for batch purposes. It is combining stream processing with batch processing that benefits from having only one code base or pipeline for entire job execution. Nevertheless, one might have to deal with separate executions and inconsistencies between batch and streaming if Lambda architecture would be used. Moreover, Samza employs a reactive mechanism to deal with out of order message delivery. In the case of out-of-order messages, Samza finds out the partition for the message and recomputes it by replaying messages again.

Samza relying on a resource manager like YARN can scale as much as the underlying YARN cluster can. Samza has a decentralized mechanism when managing jobs. Each job has a coordinator. All jobs have a set of containers. Depending on the load, one can address the scaling of jobs independently.

6.3.1.2 Kafka Streams

Kafka Streams is a client library for building applications on top of Kafka. Kafka Streams API provides the ability to transform data in a simplified way. It does not require a separate cluster for execution. All execution happens in the application. Streams API communicates with other instances of the applications and automatically distributes the work among all instances. It uses a similar approach to Samza and keeps a local state in the running node itself for a task where it is a unit of work. Let us have a look at the anatomy of a Kafka Streams application shown in Figure 6.8.

Before diving into details of Kafka Streams, let us check out an example program where we count suspicious transaction counts for a given user in the last 15 minutes. In this example, we consume a stream from topic transactions. We filter transactions that have an amount of more than 1000. We group filtered transactions by user id and window it by 15 minutes. We then count transactions per user id using the local state. Finally, we pass down the user id to count pairs to the downstream topic. Perhaps, we can fire an alert in the downstream application or lock the user account before a further loss has been recorded.

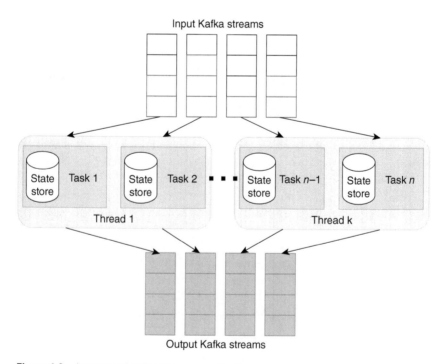

Figure 6.8   Anatomy of Kafka Streams application.

```
builder.stream("transactions",
    Consumed.with(Serdes.String(), transactionSerde))
    .filter((transactionId, transaction) -> transaction.amount > 1000)
    .groupBy((transactionId, transaction) -> transaction.userId)
    .windowedBy(TimeWindows
        .of(Duration.ofMinutes(15))
        .grace(Duration.ofMinutes(3)))
    .count(Materialized.as("transaction-count-store-15-minutes"))
    .toStream()
    .map((key, value) -> KeyValue.pair(key.key(), value))
    .to("suspicious-user-transaction-counts",
        Produced.with(Serdes.String(), Serdes.Long())));
```

Tasks Kafka keeps messages in a partitioned fashion. Kafka Streams leverages the same philosophy for distributing work through tasks. A stream partition contains an ordered sequence of records that maps to messages in Kafka topic. Kafka Streams create a fixed number of tasks for given input streams where it assigns the same partitions for a task. Tasks instantiate their processor topology and process one message at a time independent of each other. The number of stream tasks depends on the number of partitions on the Kafka topic. If there are 10 partitions, we can at most have 10 task instances running parallel. Hence, the number of topic partitions limits the parallelism of tasks.

Kafka Streams do not come with a dedicated resource manager. Kafka Streams is just a client library that can run anywhere. One can run multiple instances of the same application in the same machine or distribute the work among Kubernetes instances. Kafka Streams can distribute the work among tasks in different nodes automatically. The assignment of partitions never changes. If an application instance dies during the execution, all assigned tasks for that instance would be restarted on other instances.

Threading The number of threads in an instance is configurable for a Kafka Streams application. We can set a higher number of threads to further parallelize the work. Each thread runs one or more tasks independently. There is no shared state between the threads. This makes it quite easy to scale applications without worrying about coordination between threads.

Scaling the application is relatively easy. We can have more threads in an instance, or we can have more instances. Kafka Streams automatically assigns partitions among various stream threads through Kafka's coordination mechanism. We can start as many as application instances till we equalize the number of partitions. Once we reach the number of partitions, some of the threads would stay idle until failure occurs.

Recovery Kafka Streams rely on Kafka, which is a robust and reliable infrastructure. If processing is passed down to Kafka, it would be persisted even if the application itself fails. Moreover, if a node fails, the tasks on that node would automatically redistribute among the remaining nodes.

Kafka Streams use the local state for implementing stateful applications. It is important to have a robust local state, too. Thus, Kafka Streams keep incremental changes replicated in a Kafka topic. With a changelog, Kafka Streams can track any state updates. If a machine fails, Kafka Streams can recover the latest local state by replaying the changelog before resuming the processing. Nevertheless, the task initialization in case of failures might still take some time. For fast recovery, it might be good to set standby nodes so that recovery happens faster.

6.3.2 Apache Pulsar

Apache Pulsar is an open-source, distributed, and horizontally scalable publish/subscribe messaging solution. Pulsar has the concept of topics similar to Kafka. Producers publish to a topic. Consumers subscribe to a topic and process messages. Pulsar will retain messages even if the consumer gets disconnected. The messages would be ready when the consumer reconnects. Pulsar discards messages once the consumer acknowledges them (Apache Pulsar, 2020). Apache Pulsar supports some of the key features for messaging as follows:

- Pulsar can support multiple clusters over multiple geolocations. Pulsar replicates messages between geolocations seamlessly.
- Pulsar has API clients for popular languages such as Java, Python, Go, and C++.
- Pulsar provides different modes of subscription to the topics, covering many use cases.
- Pulsar persists messages in a tiered fashion where it can offload data from warm storage to cold storage. Pulsar can restore data from cold storage without clients noticing it.
- Pulsar offers a streaming computation API; Pulsar functions, to accommodate native stream processing capabilities. Pulsar functions can make use of Pulsar IO, making it easy to move data in and out.

Messaging Overview

Pulsar groups topics into logical namespaces. A namespace provides isolation of different applications as well as a hierarchy between topics. Moreover, Pulsar further categorizes topics through the concept of tenants. Tenant is a core concept in Pulsar to embrace multi-tenancy. Tenants are administrative units that facilitate configuring storage quotas, isolation policies, authentication schemes, and more. Hence, a topic URL consists of persistence, tenant, namespace, and topic as follows:

persistent://tenant/namespace/topic

Topics can be partitioned. By default, Pulsar serves topics through one broker. Nevertheless, this limits throughput to the maximum throughput of one broker.

Thus, Pulsar provides partitioned topics. Internally, partitioned topics are simply n individual topics, where n is the number of partitions. Pulsar automatically distributes the messages to the right partitions. When using partitioned topics, we need a routing mode to define which partition a message goes.

Pulsar has different subscription modes to the topics: exclusive, shared, and failover illustrated in Figure 6.9. In exclusive mode, only one consumer can process messages. In failover mode, one consumer is the master and processes messages, whereas others are inactive until the master disconnects. Finally, in shared mode, consumers receive messages in a round-robin fashion. The shared mode has another submode called, key shared where Pulsar guarantees delivering the same messages to the same consumers based on key. Moreover, Pulsar allows multiple subscriptions for a consumer. A consumer can subscribe to multiple topics with the help of regular expressions. For example, a consumer can subscribe to all topics about reviews by *persistent://public/default/reviews-.**.

When it comes to storing messages, Pulsar offers persistent and nonpersistent topics. Persistent topics can survive broker restarts and subscriber failovers. Nevertheless, nonpersistent can't survive failures since all messages are kept in memory. Nonpersistent is part of the topic URL. When Pulsar persists messages, it applies a retention policy and expiry. Message retention provides storing messages after a consumer acknowledges it. On the flip side, expiry determines a time to live (TTL) for messages before they are acknowledged. Besides, Pulsar can apply deduplication to messages so that the same message gets delivered only once within the retention period.

Architecture Overview

A Pulsar instance consists of one or more Pulsar clusters. Clusters can replicate data with each other. Each cluster consists of three major components: Pulsar brokers, a ZooKeeper cluster, and a BookKeeper cluster. Pulsar brokers accept messages from producers and dispatches messages to the consumers based on

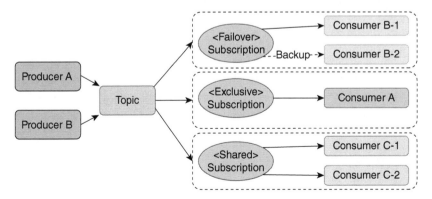

Figure 6.9 Pulsar topic subscription modes.

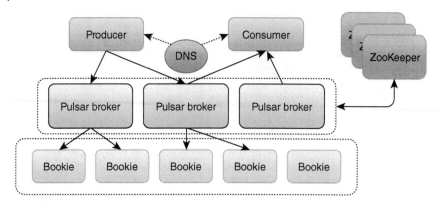

Figure 6.10 Pulsar architecture.

the subscription mode. The BookKeeper cluster is responsible for persisting messages. The ZooKeeper cluster is responsible for the coordination of various tasks. Figure 6.10 shows an overview of Pulsar's architecture.

Pulsar brokers consist of a dispatcher and an HTTP server. The dispatcher routes messages from producers to consumers, while the HTTP server provides an administration interface and topic lookups for producers and consumers. Pulsar brokers are stateless. They do not store any information locally other than the cache for messages. Finally, Pulsar brokers provide geo-replication for global topics. They republish messages that are published in the local region to remote regions.

Pulsar uses ZooKeeper for metadata storage. It keeps global configuration like tenants, namespaces, and other entities that need to be consistent globally. Each cluster has its ZooKeeper cluster. In each local ZooKeeper cluster, Pulsar stores information like ownership, BookKeeper ledgers, load reports, and so forth.

Pulsar guarantees persistence for messages. If any consumer does not acknowledge a message, it will stay in Pulsar until expiry. Pulsar uses BookKeeper for reliable and low-latency message storage. BookKeeper is a distributed write ahead log (WAL) system that has a base unit called ledger. A ledger is an append-only sequence of entries. Pulsar utilizes ledgers to write many independent logs. Pulsar creates multiple ledgers for a topic over time. BookKeeper offers read consistency over a ledger. BookKeeper distributes ledgers of clusters of bookies where a bookie is a node in the BookKeeper cluster. Furthermore, BookKeeper offers even distribution of load over bookies and horizontally scalable. Pulsar keeps cursor information in BookKeeper, too. BookKeeper enables Pulsar to recover the cursor position in case of failures.

6.3.2.1 Pulsar Functions
Pulsar functions are lightweight computing facilities to apply transformations on streams. Pulsar functions consume messages from multiple topics, apply the

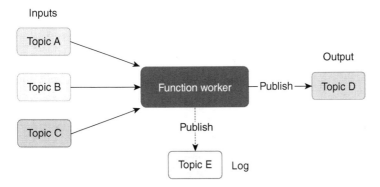

Inputs

Topic A

Topic B

Topic C

Function worker

Publish

Output

Topic D

Publish

Topic E   Log

Figure 6.11 Pulsar functions programming model.

computational logic in the function, and publish the results to another topic. Whenever a message arrives, the function will write the result to an output topic and BookKeeper. It can also write logs to a log topic so that potential problems can be detected. Pulsar supports functions API for Java, Python, and Go. Functions can run inside a broker or outside of a broker. The programming model for Pulsar functions is shown in Figure 6.11. Pulsar functions are very simple. They use a language native interface to expose the logic. They do not depend on any Pulsar-specific libraries. To write a Pulsar function, we simply extend *Function* interface of Java or create a function in Python or Go. Nevertheless, we might need to use Pulsar-specific functionalities like state management or configuration. Let us see a Java example of two Pulsar functions. The first example does not depend on Pulsar libraries. It simply reverses a given string and publishes it back to the output topic.

```
public class ReverseFunction implements Function<String, String> {
  public String apply(final String input) {
    return new StringBuilder(input).reverse().toString();
  }
}
```

The second example uses Pulsar libraries, specifically *Context* variable, to keep track of the state. It calculates averages by keeping count and previously calculated average in the *Context*. Context provides contextual information to the executing function. It also allows functions to publish to different topics.

```
public class AverageFunction
    implements Function<Integer, Double> {
  @Override
  public Double process(final Integer input,
        final Context context) {
    context.incrCounter("score-count", 1);
    final long currentCount = context.getCounter("score-count");
```

```
      final double currentAvg = Optional
          .ofNullable(context.getState("score-avg"))
          .map(byteBuffer -> toDouble(byteBuffer)).orElse(0.0);
      final double newAvg =
          (currentAvg * (currentCount - 1) + input) / currentCount;
      context.putState("score-avg", toByteBuffer(newAvg));
      return newAvg;
  }
}
```

Pulsar allows windowing through configuration parameters. There are two common ways to define windowing logic. Pulsar allows us to define window length and sliding interval in terms of count and time. Window length determines when we go to the next window, e.g. every five minutes. Sliding interval determines duration or count after which the window slides. Implementing windowing is easy. We just have to accept a collection of messages instead of a single message. Let us have a quick look at the following example, where we would listen to sensor readings for the last five minutes. If the temperature is higher than the threshold, we will publish to a sensor alarm topic. Note that we get the threshold through *WindowContext* config. Again, we publish with the help of context. We configure the window by setting *windowLengthDurationMs* to five minutes.

```
public class AlarmFunction
    implements WindowFunction<SensorReading, Void> {
    @Override
    public Void process(Collection<Record<SensorReading>> input,
        WindowContext context) {
      final double temperatureThreshold = Optional
          .ofNullable(context.getUserConfigMap()
          .get("temperatureThreshold"))
          .map(o -> (double) o).orElse(100.0);
      final Map<String, SensorReading> maxTemperatures =
          input.stream()
          .map(sensorReadingRecord ->
            sensorReadingRecord.getMessage())
          .filter(Optional::isPresent)
          .map(sensorReadingMessage ->
            sensorReadingMessage.get().getValue())
          .filter(reading ->
            reading.getTemperature() > temperatureThreshold)
          .collect(toMap(SensorReading::getSensorId, identity(),
              maxBy(Comparator
                .comparing(SensorReading::getTemperature))));
      for (SensorReading reading : maxTemperatures.values()) {
        context.publish("sensor-alarms",
```

```
        new SensorAlarm(reading));
    }
    return null;
  }
}
```

Pulsar functions support three ways to run the functions: thread, process, and Kubernetes. In thread mode, Pulsar invokes functions in threads. In process mode, Pulsar invokes functions in processes. In Kubernetes mode, Pulsar submits functions to Kubernetes stateful set. When running in thread or process mode, functions can run within the broker. Nevertheless, this might not be sustainable if the amount of work for functions gradually increases. It is better to run Function Worker in a separate cluster, especially if we have a Kubernetes cluster. We can potentially use multiple proxies. We can configure one or more proxy for Pulsar Function admin requests and isolate function workload from broker workload as shown in Figure 6.12.

6.3.3 AMQP Based Brokers

Before the advent of systems like Kafka and Pulsar, AMQP-based message brokers were popular and addressed many use cases. Some of the popular AQMP implementations are RabbitmMQ, ActiveMQ, Apache QPid, and StormMQ. We cannot do stream processing with these message brokers. They do not offer an execution engine to do further processing. Nevertheless, we can use a stream processing engine to process messages as part of a stream through connectors. In this case, the broker would route the messages as usual. And, a stream engine will consume

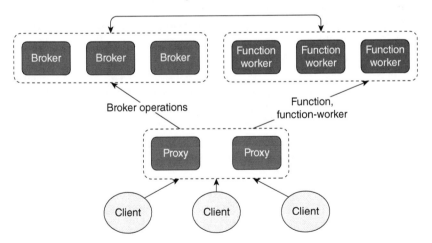

Figure 6.12 Pulsar functions worker.

messages as one of the consumers. The connectors usually implement a couple of approaches to steam data from AMQP implementation to the streaming engine. Some approaches consume messages through one worker, and others consume through many workers. The partitioning of data and persistence depends on the streaming engine.

6.4 Streams via Stream Engines

Stream processing engines are specially built for stream processing. Stream processing engines provide distributed, scalable, and fault-tolerant scaffolding for users to build streaming applications. In many cases, stream processing engines are agnostic about data sources. They have a pluggable interface for accepting sources through connectors. A stream processing engine might bring many additional capabilities for a modern Big Data platform. This section will go through widely adopted stream processing engines such as Flink, Storm, Heron, and Spark Streaming.

6.4.1 Apache Flink

We started our discussion on Apache Flink in Chapter 5. We mentioned Flink handles batch processing as a special case of stream processing. Flink accepts data streams from various resources such as message queues, socket streams, and file systems through connectors. Then, Flink applies transformations on the stream and writes output results via sinks. Flink is very flexible when it comes to execution. It can run inside other programs, standalone, or on a cluster of machines. In this section, we will talk about how Flink handles stream processing and describe the components of Flink.

Flink can create streams from a collection of items, file systems, and sockets. Nevertheless, it is more common and pervasive to create streams from a messaging layer. We can choose a messaging layer as Kafka, Pulsar, and more. Flink returns results of processing via sinks. Sinks can be various systems such as Cassandra, HDFS, and so forth. Let us now see an example of stream processing where our Flink application would receive call detail records (CDR). Our Flink application will write to another Kafka topic to notify the customer if their usage passes their customer-set threshold.

```
final StreamExecutionEnvironment env =
   StreamExecutionEnvironment.getExecutionEnvironment();
env.setStreamTimeCharacteristic(TimeCharacteristic.EventTime);
env.getCheckpointConfig()
   .setCheckpointingMode(CheckpointingMode.EXACTLY_ONCE);
```

```
env.getCheckpointConfig()
  .setCheckpointInterval(TimeUnit.MINUTES.toMillis(10));
env.setStateBackend(new FsStateBackend
  ("hdfs://dbdp/flink/checkpoints/"));

//We pass down properties to get consumer and producer
final SourceFunction<CDR> source = getConsumer(properties);
final SinkFunction<OverUser> sink = getProducer(properties);
final DataStream<CDR> messageStream = env.addSource(source)
    .assignTimestampsAndWatermarks(new CDRTimestampAssigner());

messageStream
  .map(value -> Tuple2.of(value.phoneNumber, value.end - value.start),
      TupleTypeInfo.getBasicAndBasicValueTupleTypeInfo
      (String.class, Long.class))
  .setParallelism(10) //Sets the parallelism for map operator.
  .keyBy(0)
  .window(TumblingEventTimeWindows.of(Time.hours(2)))
  .allowedLateness(Time.minutes(10))
  .reduce((value1, value2)
    -> Tuple2.of(value1.f0, value1.f1 + value2.f1))
  .filter(value -> getThreshold(value.f0) < (value.f1))
  .map(value -> new OverUser(value.f0, value.f1))
  .addSink(sink);
```

In our CDR example, we create a source and sink from Kafka. We then map CDR values to a pair of phone numbers and call duration. We group calls by phone number and set a window size of half an hour. Later, we sum up the duration by reducing each pair into one. Once we have a total duration, we filter them by the customer-set threshold. We call an external function for threshold, typically, a cache call backed by the database. We convert our tuple into a concrete object and pass it to a sink in the last step.

Let us see how Flink executes applications under the hoot.

6.4.1.1 Flink Architecture

Flink has a layered architecture. At its core, it has a runtime where it receives a job graph to execute. A job graph represents a flow of transformations applied to incoming data streams and publishing results to an output data stream. Underneath the runtime, Flink supports a variety of deployment models where the actual execution happens. On top of runtime, Flink comes with libraries to assist in processing data streams. As of now, Flink has two separate sets of libraries to address stream and batch processing. Nonetheless, the Flink community has a long-term plan to unify the libraries into one. The top layer provides an extended API for

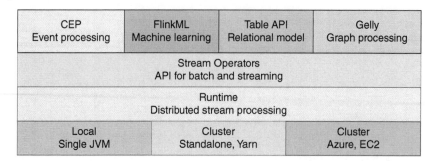

CEP Event processing	FlinkML Machine learning	Table API Relational model	Gelly Graph processing
Stream Operators API for batch and streaming			
Runtime Distributed stream processing			
Local Single JVM	Cluster Standalone, Yarn		Cluster Azure, EC2

Figure 6.13 Flink architecture.

relational table querying, machine learning, graph processing (Flink, 2020). The flink architecture is shown in Figure 6.13.

Flink architecture allows running computations over bounded and unbounded streams. Unbounded streams have no defined end. Order plays a significant role in unbounded stream processing since data still flows through the engine. Flink supports precise control over time and state for unbounded streams. On the flip side, bounded streams have a defined end. Order is not an issue for a bounded stream since it can be sorted. Flink uses data structure and algorithms specifically designed for bounded streams to get the best performance.

Flink processes both bounded and unbounded streams by a basic unit called a task. Tasks execute a parallel instance of an operator. Thus, an operator's life cycle also defines the life cycle of a task. An operator goes through several phases. In the initialization phase, Flink prepares the operator to run, sets the state. In the process element phase, the operator applies the transformation to the received element. In the process watermark phase, the operator receives a watermark to advance the internal timestamp. In the checkpoint phase, the operator writes the current state to external state storage asynchronously. Tasks contain multiple chained operators. In the initialization phase, tasks provide the state of all operators in the chain. The task's wide state helps with recovering from failures. After setting state, the task runs the chain of operators until it comes to an end for bounded streams or canceled for unbounded streams.

Flink can run tasks in many different environments. Flink integrates with all common clustering technologies such as YARN, Mesos, or Kubernetes. It can also run as a standalone cluster or local JVM for development purposes. Flink provides resource manager-specific deployment options. Flink automatically defines the required resources for processing streaming applications. It then negotiates resources with the resource manager. Flink replaces failed containers with new ones.

6.4.1.2 System Design

Flink processes streaming data seamlessly with the help of some of the critical system design choices such as fault-tolerance, job scheduling, and abstractions over persistence.

Fault Tolerance Flink provides a fault tolerance mechanism to recover the state consistently. The fault tolerance mechanism allows Flink to generate snapshots of streaming flow. In case of a failure, Flink stops the flow and resets streams to the snapshot and operators to the checkpoint. Flink produces snapshots with the concept of barriers. Barriers are signs that separate the records from one snapshot to another. They flow with the data stream, as shown in Figure 6.14. When multiple barriers are in the stream, it means various snapshots happen concurrently. The barriers contain information about the last record. For example, it is the partition offset of the last record for Kafka.

When an operator receives barriers from all of its input streams, it emits the barrier for all output streams. When the sink operator at the end of the stream receives a barrier, it acknowledges the snapshot to the checkpoint coordinator. Once the job processes snapshot, it would not ask for records from the streams. If there is more than one input stream, the operator has to align the input streams. The operator will put the records to a buffer up until it receives all barriers from other streams. This is required since the operator cannot mix records from one barrier to another.

Since the alignment process needs to wait for each barrier to arrive, it can take some time. Flink supports skipping alignment in case of checkpoints. Nevertheless, this can result in processing the same data more than once. If the alignment is skipped, the operator will process all inputs. At some point, all barriers will arrive; however, messages outside of the current barrier, belong to the newer barrier, would be already computed. If a failure happens, the operator will receive messages that belong to the new barrier for recovery.

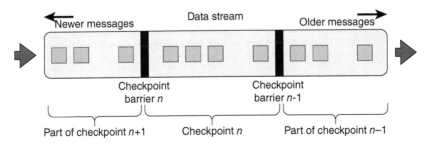

Figure 6.14 Flink barriers.

Nevertheless, some of the messages were already computed since alignment was skipped. Consequently, it will result in executing the same messages multiple times. We can configure this behavior by setting exactly once or at least once for checkpoints.

Other than the positions of the streams, Flink saves a snapshot of user-defined states and system state. The user-defined state is the state that keeps user transformations. The system state keeps data buffers, which are part of the computation. The snapshot might be large since they contain the position of each stream before the snapshot happened, and the state for each operator. Hence, Flink provides mechanisms to save the snapshot to an external backend like HDFS. Saving snapshots can happen both synchronously and asynchronously, depending on the operator. Asynchronous snapshots are possible when updates to the state do not affect that state object.

Scheduling A Flink job is a pipeline of operators where we can set parallelism on the operator level. Flink executes operators with tasks. Each execution resource has task slots which define how many tasks that it can execute. Execution resources control task allocation through task manager (task executors). For example, if we have three execution resources with three task slots and an operator that requires parallelism of 7, the program will distribute tasks over resources, as shown in Figure 6.15.

Before starting an execution, Flink creates a job graph. A job graph is a graph of vertices, and edges form a DAG. A vertex in the graph represents an operator. An edge in the graph represents intermediate results. Each vertex has an application logic and extra libraries and resources required to run the operator. At execution time, the job manager (job master) transforms the job graph into an execution graph. The execution graph is the parallel version of the job graph. The execution graph has an execution vertex per job vertex where each execution vertex keeps track of a task. If an operator has a parallelism of 100, then it would have 100 execution vertices. Execution graph keeps track of global operator state

Figure 6.15   Flink task scheduling.

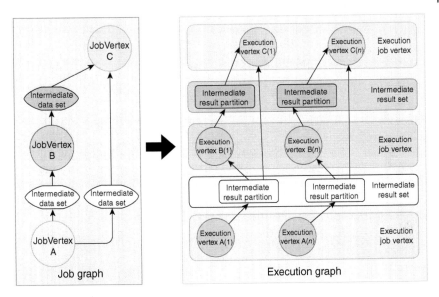

Figure 6.16 Execution graph.

through execution job vertex. Apart from vertices, the execution graph contains edges that contain intermediate results that contain intermediate result partitions. The execution graph is illustrated in Figure 6.16.

A job goes into several states during the execution. Job starts with creation state, transitions to scheduling, and then running and finally end up with a finished, canceled, and failed state. If all vertices have finished successfully, the job transitions to a finished state. In case of failures, the job transitions to a canceling state where all running tasks are canceled. If the job is not restartable, then the job transitions to failed. If the job can be restarted, then it will enter the restarting state and then be created.

Persistence Abstraction   Flink uses file system abstraction, which has common file system operations with minimal guarantees, e.g. consistency. The operations in the file system are limited. Updates to existing files are not supported. File system abstraction provides integration with common locations such as HDFS, object store, and the local file system.

6.4.2   Apache Storm

Apache Storm has emerged as one of the earlier systems for real-time stream processing and has been widely adopted. Storm provides primitives to process a huge set of data at scale. Storm primitives are fault-tolerant and have data

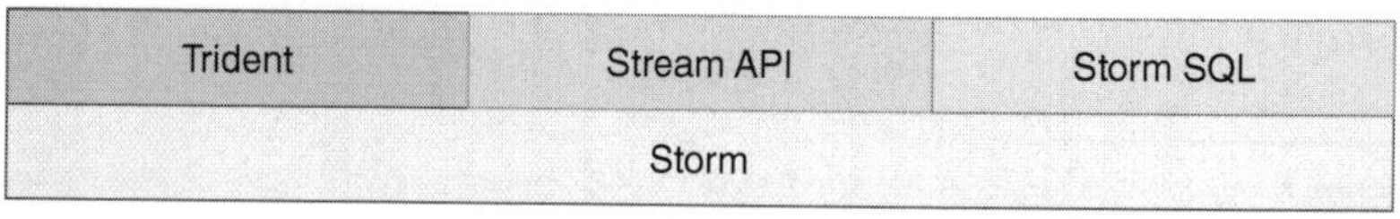

Trident	Stream API	Storm SQL
Storm		

Figure 6.17 Storm layers.

processing guarantees. Storm primitives integrate with many different data technologies such as HDFS, Kafka, Cassandra, and more. On top of primitives, Storm has layers to provide additional functionality. Trident, an alternative interface, provides exactly-once processing and better persistence guarantees. Stream API provides a wide range of transformations such as mapping, filtering, aggregations, and windowing. Storm SQL empowers users to run SQL queries over streaming data (Apache Storm, 2019) (Figure 6.17).

Storm consists of tuples and streams. A tuple is an ordered list of named fields. A stream is an unbounded sequence of tuples flowing from a source to one or many destinations. Tuples can have dynamic types. Storm can serialize all primitive types. Though, we need to register the serialization of complex types to configuration. Streams of tuples flow through a Storm program with the help of spouts and bolts.

A Storm program forms a topology that is a DAG of spouts and bolts connected with stream groupings. Spouts are sources of streams. Usually, spouts read tuples from an external source and emit them into the topology. Spouts can be reliable and unreliable. If a spout is reliable, it can replay a tuple that the program could not process. Nonetheless, unreliable spouts forget about tuples once they are emitted; tuple processing fails when it is unreliable.

Bolts are processing units. Bolts can transform streams in many ways through filtering, aggregations, joins, windowing, and so forth. Bolts can accept more than one stream. At the end of the transformation, bolts can emit values with their respective output collectors. On the flip side, bolts can write data into another system and finish the topology. When setting bolts, we define stream groupings. Stream groupings specify which streams bolt should receive. Besides, stream groupings define the partitioning of streams between bolt tasks. Figure 6.18 shows an example Storm flow.

We will go over an example Storm program that makes a very basic sentiment analysis on brands by reading tweets. The Storm flow has a spout which reads the tweets and emits them. Sentiment bolt picks up tweets and transforms them into a brand, sentiment, and timestamp tuple. Weighted sentiment bolt picks up tuples in a window of 10 minutes and decides the weighted sentiment for the brand with start and end. In the last step, we persist weighted sentiment into durable storage. Note that we define parallelism for each bolt by setting the number of tasks in the builder. Our topology runner would like as follows:

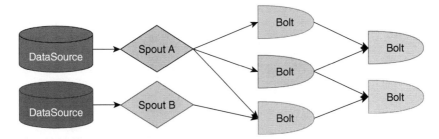

Figure 6.18 Storm spouts and bolts.

```
final TopologyBuilder builder = new TopologyBuilder();
//Receives tweets and emits
builder.setSpout("tweetSpout", new TweetSpout());
//Determines brand and sentiment
builder.setBolt("sentimentBolt", new SentimentBolt())
    .shuffleGrouping("tweetSpout").setNumTasks(10);
//Finds weighted sentiment for a brand in last 10 minutes
builder.setBolt("weightedSentimentBolt", new WeightedSentimentBolt()
    .withTimestampField("createdAt")
    .withLag(BaseWindowedBolt.Duration.minutes(1))
    .withWindow(BaseWindowedBolt.Duration.minutes(10)))
    .partialKeyGrouping("sentimentBolt", new Fields("brand"))
    .setNumTasks(3);
//Persists the results into a storage
builder.setBolt("persistingBolt", new PersistingBolt())
    .shuffleGrouping("weightedSentimentBolt");

final Config config = new Config();
config.setDebug(false);
final ILocalCluster cluster = getCluster();
cluster.submitTopology("tweets", config, builder.createTopology());
```

Our weighted sentiment bolt is special since it has windowing for the last 10 minutes. Storm will group tuples by the brand since we specify it in our topology. Storm applies a hash function on the field or fields. So, a single processor can receive multiple brands. Weighted sentiment bolt simply counts sentiments and choose the sentiment with a maximum number of occurrences as the weighted sentiment.

```
public void execute(final TupleWindow tupleWindow) {
    final List<Tuple> tuples = tupleWindow.get();
    if (tuples.isEmpty()) {
        return;
    }
    tuples.sort(comparing(s -> s.getLongByField("createdAt")));
    final long beginningTimestamp = tuples.get(0).
        getLongByField("createdAt");
```

```
final long endTimestamp = tuples.get(tuples.size() - 1).
    getLongByField("createdAt");
tuples.stream()
    .collect(groupingBy(v -> v.getStringByField("brand"),
        collectingAndThen(
            groupingBy(v -> v.getStringByField("sentiment"),
                counting()),
            sentimentCounts -> sentimentCounts.entrySet().
                stream()
                    .max(comparingByValue()).get().getKey()
    )))
    .forEach((brand, weightedSentiment) -> outputCollector
        .emit(new Values(brand, weightedSentiment,
            beginningTimestamp, endTimestamp)));
}
```

Now that we have seen Storm in action, we can discuss how Storm executes the programs.

6.4.2.1 Storm Architecture

Apache Storm runs topologies in a distributed fashion. Since topologies process streams, they never finish. A common scenario is to save the results of the topology to persistent storage. Storm runs topologies with a master node, called Nimbus, and worker nodes, called Supervisors. The architecture overview is shown in Figure 6.19.

Master node, similar to coordinators from other distributed Big Data solutions, keeps track of the state of topologies. The master node has a daemon called

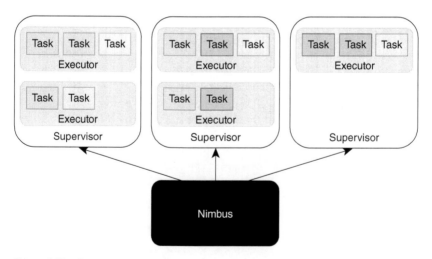

Figure 6.19 Storm architecture.

Nimbus to keep track of topologies. Nimbus distributes application code among worker nodes, assigns tasks, monitors failures, and restarts failed tasks whenever it is possible. Nimbus keeps the state in ZooKeeper and can be configured as active/passive where the passive node becomes active when the active fails. Change of active node does not disrupt much since the state is in ZooKeeper. If the Nimbus goes down, workers will continue working, but no new jobs can be accepted as there is no master node.

Worker nodes run a supervisor daemon, which creates, starts, and stops worker processes. Supervisor nodes also keep their state in ZooKeeper; therefore, restarting supervisor nodes is smooth. A worker process executes a subset of topology. Worker processes spawn threads to run one or more executors to run spouts and bolts in the topology. Each executor runs one or more tasks where each task process data for a spout or bolt. The number of tasks is always the same; however, the number of executors can change. Hence, the number of tasks is greater or equal to the number of executors. Storm tries to run one task per executor by default.

6.4.2.2 System Design
Storm has several system design choices, which makes it a reliable and performant stream processor. Two key elements are scheduler and message processing guarantees.

Scheduler Storm has a pluggable scheduler. One can write their scheduler by simply implementing the scheduler interface. There are two popular schedulers: isolation and resource-aware scheduler.

The isolation scheduler isolates running topologies from each other in a storm cluster. The scheduler gives the option to specify which topologies should be isolated. Isolation simply means running specified topologies on dedicated hardware. Isolated topologies take priority if they are to compete with non-isolated topologies. The scheduler makes sure isolated topologies get the resources they need. Once all isolated topologies get resources, the remaining resources will be shared among other topologies.

The resource-aware scheduler provides multitenancy since a Storm cluster is generally shared among many users. The scheduler can allocate resources based on the user. The scheduler tries to run each user's topology whenever possible. When there are additional resources in the cluster, the scheduler distributes resources to users fairly. Users can set priority for their topology. The scheduler will consider when allocating resources or evicting topologies when there are not enough resources.

Reliability Storms have different types of guaranteed message processing. It can do at-most-once, at-least-once, and exactly-once processing with the help of Trident.

Storm marks a tuple as failed if one or more succeeding tuples fail in the tuple graph formed by bolts and spouts. Subsequent tuples can fail due to runtime errors or timeouts. Storm has special tasks called ackers that keep track of every tuple in a DAG when enabled. Thus, when a failure happens, Storm can report it to the spout that produced the tuple.

To support at-most-once, nothing has to be done. Storm drops tuples if they fail when they go through the DAG. For at least once, Storm tracks each tuple until they reach the end before timeout. If a tuple fails along the way, Storm reemits the tuple into the DAG. Nevertheless, this can cause reprocessing of the same tuple more than once. For a better processing guarantee, we can use Trident, which supports exactly once semantics.

The most common scenario is to enable at-least-once processing. There are a couple of things that need to happen for at-least-once.

- Ackers need to be enabled.
- Spouts need to specify a unique id for each message. This comes out of the box from spout implementations like Kafka.
- Bolts need to anchor tuples that they emit, with input tuples.
- Bolts need to acknowledge a tuple when they are done processing.

Storm implements a pretty cool algorithm based on XOR to keep track of tuples without exhausting memory.

6.4.3   Apache Heron

Apache Heron is the successor of Apache Storm with a backward compatible API. Heron surpasses Storm in a couple of areas such as debugging, multitenancy, performance, scaling, and cluster management. Besides, Storm has a couple of limitations when running on production. One of the key challenges has been debugging. It is quite hard to determine performance degradation on Storm topology. Provisioning is also a hard problem since Storm needs dedicated resources. What is more, allocating/deallocating resources requires manual work (Kulkarni et al., 2015).

6.4.3.1   Storm Limitations
Storm workers run different tasks in the same machine and log them to the same file. A Kafka spout and a bolt independent of each can run in the same worker. Since they have different characteristics, it is hard to detect misbehaving topologies or steps in a flow. Furthermore, problems from one topology can affect others since they can share the same worker. At this point, resource allocation becomes another problem.

Storm needs a homogeneous environment where each worker has the same type of resources. When tasks require disparate resources like memory, the maximum required resource need to be selected for all components. For instance, let us take a topology where we have three bolts and five spouts where bolts require 10 GB memory and spouts require 5 GB memory on three workers. Each worker needs 20 GB resulting in 60 GB in total, although 55 GB would be sufficient. Thus, such issues cause inefficiencies, result in wasting resources, and require oversubscribing on the infrastructure.

Storm's Nimbus has too many responsibilities to address. It has to do the distribution of topology, scheduling, monitoring, and reporting. At times, Nimbus can get overwhelmed and fail. Nimbus allows running different topologies in the same machine where topologies can affect each other negatively. If one of the topologies is hosed, it might end up shutting down the whole cluster.

Lastly, Storm does not offer any solutions for back pressure. Storm simply drops messages. Nevertheless, this is rarely what we want. All work happened before the overloaded step gets lost.

6.4.3.2 Heron Architecture

Heron relies on well-known schedulers like YARN or Mesos as oppose to Storm's Nimbus use. Moreover, Heron addresses complications and problems in Storm's design by leveraging divide and conquer responsibility in terms of responsibility. The architecture diagram is shown in Figure 6.20. When a job is submitted, Heron creates containers for topology masters and several worker containers where each one has a stream manager, metrics manager, and multiple Heron instances.

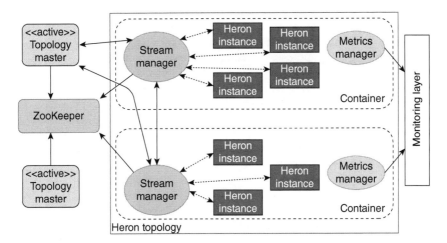

Figure 6.20 Heron architecture.

Topology Master Topology Master manages topology. It provides application status and metrics. At any time, there can be only one topology master. The active/standby coordination happens through ZooKeeper.

Stream Manager Stream manager is the gateway for tuple exchange. All stream managers connect. Each Heron Instance connects to their local stream managers. Heron employs back pressure mechanisms through stream managers by clamping down the spouts so that new data would not be injected into the topology.

Heron Instance Heron Instance is where the actual work happens. Each Heron Instance only runs a single spout or bolt task since it makes it easier to debug or monitor. Heron operates two threads per Heron Instance. The first thread, gateway, is responsible for communication with Stream Manager. The second thread, task execution, is responsible for the actual execution of tasks. Since all communication happens through stream managers, it is possible to have Heron Instances in different languages other than Java.

Metrics Manager The metrics manager collects and reports metrics to the metrics layer. Metrics include system metrics, user metrics, and topology metrics.

Heron Tracker Heron Tracker is a portal to serve the information about topologies. It connects to the same ZooKeeper cluster so that it can serve the same information. Tracker watches ZooKeeper learn about new topologies or updates in topologies. The tracker provides a Rest API to make information available to multiple parties. The Rest API serves information about logical and physical plans for topologies, metrics about the system and topologies, and links to log files.

Heron UI Heron UI uses Heron Tracker to visualize information about the cluster. Users can drill down to various metrics as well as topology plans.

Heron Viz Heron Viz is a dashboarding tool that generates dashboards out of metrics collected by the metrics manager. The dashboards include information about health metrics, resource metrics, component metrics, and stream manager metrics.

6.4.4 Spark Streaming

Spark Streaming is a layer on top of Spark core API. It enables horizontally scalable, high throughput stream processing. Spark streaming stream data from message brokers like Kafka as well as other streaming data sources. Spark streaming can apply regular transformation functions that it offers in core API to the streams. Spark streaming allows saving stream results to various technologies such as HDFS and Cassandra.

Internally, Spark steaming divides data into micro-batches and produce micro-results out of these batches. Spark streaming provides a high-level abstraction called discretized streams. A discretized stream (DStream) is a continuous sequence of RDDs representing an unbounded stream of data. Each DStream periodically generates an RDD either from a stream source or transformation of an RDD.

Let us see Spark Streaming in action with the following example where we create a stream out of two Kafka topics that carry a cell trace records (CTR) of base stations. Each CTR contains a series of events. Our program takes CTR stream, groups CTRs by their region id, filters events by call drop event type, and counts the events. Later, the program filters down the call drop count if a region exceeds a certain threshold. At the last step, the program writes results to Cassandra with a TTL for operational teams to conduct a deeper analysis.

```
val topics = Array("ctr-3G-radio", "ctr-4G-radio")
val stream = KafkaUtils.createDirectStream[String, CTR](
  streamingContext,
  PreferConsistent,
  Subscribe[String, CTR](topics, kafkaParams)
)
stream.map(record => (record.key, record.value))
  .map(pair => (pair._2.regionId, pair._2.events
    .count(c => c.eventType == CellEventType.CALL_DROP)))
  .window(Minutes(5))
  .groupByKey()
  .map(t => (t._1, t._2.sum))
  .filter(t => t._2 > DROP_COUNT_THRESHOLD)
  .foreachRDD(rdd => {
    rdd.saveToCassandra("cell_reporting", "region_cell_drop_count",
      writeConf = WriteConf(ttl = TTLOption.constant(60 * 60)),
      columns = SomeColumns("region_id", "drop_count"))
  })
```

6.4.4.1 Discretized Streams

Spark Streaming provides DStream as the base abstraction. Each connected stream data source implements the DStream interface. DStream contains RDDs that correspond to a micro-batch. Each micro-batch is for a different interval and contains all data for the interval. Spark buffers the data until the end of the interval. The processing happens after an interval is complete. Windowing groups multiple micro-batches into a longer interval. Thus, the window size must be multiples of batch size. Otherwise, the window will end in between a micro-batch.

Spark translates operations on DStream to the underlying RDDs. For example, the mapping and filtering operation for CTR will be applied to each RDD, micro-batch, as shown in Figure 6.21. Furthermore, Spark processes micro-batches

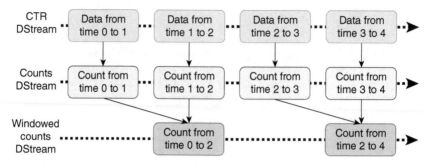

Figure 6.21   Spark streaming micro-batches.

sequentially. If the processing takes more than the batch duration, the incoming data will be stored until it is complete. Therefore, it is important to complete the micro-batch processing within a batch interval. Otherwise, delays can occur in stream processing.

6.4.4.2   Fault-tolerance

Classical Spark applications work on fault-tolerant systems; therefore, the data generated in one stage can be traced back to the source. In case of failure in a worker, the result can be recomputed by the lineage. Nevertheless, this is not the same for streaming since data sources may not provide fault-tolerant data. To achieve fault-tolerance, Spark replicates the data between the workers. In case the data is not replicated yet, then the only option is to get the data from the source if available. For some of the technologies like Kafka, it is possible to receive the data via offset when checkpoints are enabled.

A streaming application needs checkpoints to store the state in a fault-tolerant storage system to recover from failures. There are two types of data checkpointing. The first one is metadata checkpointing, where Spark saves enough information about configuration, incomplete batches, and operations. The second one is data checkpointing, where Spark saves RDDs to a reliable storage system such as HDFS. Note that storing data to reliable storage can be necessary for stateful computations.

7

Data Analytics

After reading this chapter, you should be able to

- Learn data ingestion process
- Transfer large data sets
- Apply methods to make data meaningful
- Reason about detecting anomalies
- Explain visualizing data

Big Data analytics explores a vast amount of data to uncover patterns, insights, correlations within data. Big Data analytics gives organizations opportunities and visibility of the organization. It can drive new business ideas, cost reduction, customer engagement, and better decision-making. Thus, Big Data analytics is one of the essential parts of a modern Big Data platform. Big Data analytics involves many steps such as data ingestion, data cleansing, data transferring, data transforming, data consolidation, data scheduling, dependency management, and anomaly detection. In this chapter, we will visit each one of these Big Data analytics topics.

7.1   Log Collection

One of the fundamental sources of data analytics is log collection. Many systems, such as web servers, IoT devices, and applications, generate log files. Log files can be in any form. They need to be filtered, cleansed and prepared for consumption. Before the time of microservices and containers, logging was rather straight-forward. The web servers logged files into a directory in a rotating fashion. An external component collected logs from the log directory with various sets of tools. Nonetheless, microservices with containers made things a bit more complicated. One of the key differences is the number of services. The other one is

Designing Big Data Platforms: How to Use, Deploy, and Maintain Big Data Systems,
First Edition. Yusuf Aytas.

containers. In a world of 1,000s of services with many instances, logging becomes immediately convoluted. This section will visit some of the popular logging systems that address some of these problems. Note that there are numerous logging solutions. However, we would like to keep it focused on Apache Flume and Fluentd.

7.1.1 Apache Flume

Flume is a system to collect, filter, aggregate, and transfer large amounts of log data. Flume helps to navigate logs from a source to a centralized data store like HDFS. Flume can also transfer other types of data like social media data, but we are mostly interested in log collection. Flume transfers log data as they are generated and provides streaming logs. Transferring logs immediately after they are created reduces network cost, and creates opportunities for stream processing (Apache Flume, 2019).

Flume has the concepts of events where each event is a unit of data flowing through Flume agents. An agent consists of a source and channels mapped to sinks. A Flume source receives data as an event from an external system, typically a web server. The external system must send the data in the format that the source accepts. The source publishes data into channels where a channel is a staging area that keeps events temporarily in memory or disk. The events stay in channels up until they get consumed by a sink. A sink consumes an event and writes to another source or a permanent storage system like HDFS. If the external system goes down, the agent keeps events in channels until the system comes back up. Nevertheless, if the channel capacity is reached, the agent can no longer accept events anymore. This is useful for temporary hiccups occur in the permanent storage.

Flume can consolidate events by sending them to a central agent where it accepts inputs from multiple agents and writes them to permanent storage. A typical example would be writing to HDFS. Flume can write to other technologies like Kafka through a different sink. A Flume agent installation is shown in Figure 7.1. In this example, a Flume agent simply tails the access logs through *tail-f/dbdp/access.log* and sends over to another agent in Avro format. The later agent picks up the events and saves them to HDFS via HDFS sink.

7.1.2 Fluentd

In the realm of microservices running on containers, the logging can become burdensome. Fluentd is an open-source log collector to address these challenges. Fluentd is a logging middleware that can aggregate logs from multiple resources. It formats logs into JSON objects and routes them to destinations. Fluentd integrates flawlessly with containerized microservices. It has many plugins that make it possible to move data to multiple output destinations (FluentD 2020).

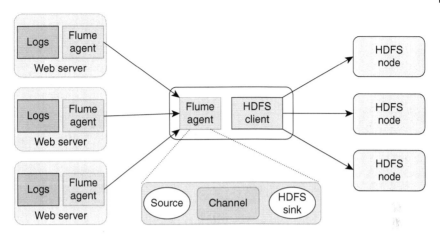

Figure 7.1 Flume agent deployment.

Fluentd treats every log as an event or a record. If the application logs four lines, then it generates four events for Fluentd. Each event consists of a timestamp and a message. An event message can be structured or unstructured. When it is structured, it conforms to JSON, where the message keys with corresponding values. Events or messages can be mutated through filters. We can drop events, partially forward, or enrich them through filters. Besides, each event gets a tag. Based on these tags, Fluentd can route events to different destinations.

Fluentd offers a buffering mechanism to overcome any issue or flakiness on downstream consumers. By default, Fluentd keeps the events in memory until they get delivered. Optionally, it can use the file system. In this case, Fluentd uses a tiered approach. When an event is ready for sending, Fluentd brings it up from the disk before sending it over to the consumers.

7.1.2.1 Data Pipeline
Fluentd is extensible in every integration point and offers a wide range of plugins supported by a big community. Each of the data pipeline components is pluggable. We can choose a plugin for each step from various plugins developed by the community. A data pipeline consists of input, parser, filter, buffer, router, and output (Figure 7.2).

Input Input is a step to collect data from logs or the environment. There are many plugins for inputs, where we can gather metrics about IO, CPU, network, and different logs.

Parser Logs come unstructured, and it would be handy if we can structure them in the source. The application still does not have to know about logging consumers.

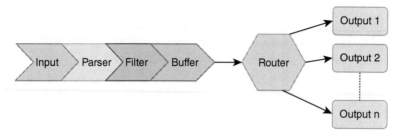

Figure 7.2 Fluentd data pipeline.

It logs entirely in the same manner. The parser takes the log and converts it into keys with corresponding values. For instance, we can process Apache Server log and get host, method, etc., parsed into JSON instead of the raw log line. There are several plugins for parsers like regex parser.

Filter The filters intercept logs. They mutate, drop, and enrich the log data with extra information. For example, attaching Kubernetes pod information to the logs can be very useful.

Buffer When logs reach buffers, it becomes immutable. Fluentd buffers logs in memory and disk, a tiered approach as we briefly discussed. Buffers give the ability to recover from network disruptions and consumer problems.

Router The router is the core mechanism to dispatch events to output destinations. The routing happens through tags and matching rules. Each record has a tag. Based on tag matching rules, the router routes messages to output locations. The matching can accept wildcards to match patterns.

Output Output plugins where the message is sent over to another system or Fluentd. There are several output plugins, such as Kafka, ElasticSearch, and more.

7.1.2.2 Fluent Bit
Fluent Bit a smaller version of Fluentd. Fluent Bit is written in C to perform well under resource-limited environments. It shares the same philosophies as Fluentd. The data pipeline is the same, but it has much fewer plugins. Fluent Bit takes 650 kB as opposed to 40 MB for Fluentd. Hence, it makes a better choice for embedded Linux environments. Moreover, it also makes a significant difference when it is deployed in containers. It takes much fewer resources so that the application can take advantage of extra memory and CPU.

7.1.2.3 Fluentd Deployment
As we mentioned previously, Fluentd is a good choice when it comes to containerized environments as it seamlessly integrates with Kubernetes and Docker. We can

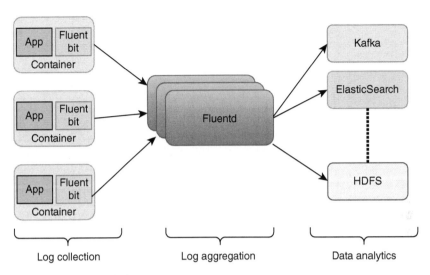

Figure 7.3 Fluentd deployment.

even set Fluentd as a log driver for Docker, which would send the logs to the configured aggregators. A practical Fluentd deployment might leverage Fluent Bit at the container and cluster of Fluentd as aggregators. Having Fluent Bit at the container saves CPU and memory and becomes less disruptive to the container application. In the Fluentd aggregators, logs can be further filtered and aggregated to send over to output destinations shown in Figure 7.3.

Fluentd supports high available deployments, delivering messages with at-most-once, at-least-once, and exactly-once semantics. In collecting a huge amount of logs without impacting the application, at-least-once and at-most-once are useful, where the data logger will be asynchronous. The log forwarders can be configured to forward logs to many aggregators or one active aggregator where the others are on standby to provide high availability. In the case of failures, the Fluentd process will be automatically started and read data from the disk buffer. Upon receiving, an event Fluentd writes them to disk buffer to avoid data loss.

Fluentd comes with a UI component, where it offers plugin installation, manages the Fluentd process, configures Fluentd settings, and views Fluentd logs.

7.2 Transferring Big Data Sets

Bringing log data and application data to the data lake is a critical part of data analytics. As discussed in the Section 7.1, we can use several tools to pump log data into our data lake. Nevertheless, we still need application data such as a user table or visitor table to make the best of the log data. Thus, we need a method to bring application data periodically to join with log data. We will look at three different elements of Big Data set transferring and tools to use.

7.2.1 Reloading

The term "reloading" means replacing the entire table in the data lake with a snapshot of the application database table. Note that the moment we start transferring the application data will change. Thus, we can never be sure what new mutations we might have. We would always assume our tables are the latest version since the start moment of the transfer. There are a couple of things to take into account while doing such transfers.

Depending on the transfer period, a metric or aggregation can change as the tables would mutate. We can anticipate some noise between hourly processing vs. daily or weekly processing. The differences can be important depending on the use case. Some of the aggregations are okay to be served after mutations. For example, aggregations for user experience. Nonetheless, some aggregations have to be recomputed as they are critical for revenue or billing.

When transferring tables, we can easily replace the table since the underlying storage system might not support transactions. We can always get caught in the middle of the switch between the old snapshot and the new snapshot. Therefore, a practical approach would be to reload tables using views. A view would point to one of the snapshots and be updated once the new snapshot is ready. Consider the following example for the hive:

```
CREATE OR REPLACE VIEW dbdp.user
    SELECT * FROM dbdp.user_ds_2020_06_30_hh_11;
DROP TABLE dbdp.user_ds_2020_06_29_hh_11;
```

We simply replace the view with the contents of the latest snapshot. We can do such reloading operations, either daily or hourly.

7.2.2 Partition Loading

When loading the whole table becomes completely impracticable, partition loading can considerably improve the performance by consuming so much computing resources. Most of the data warehouse systems generally support partitioning and make it easy to load partition instead of the entire table. A typical partitioning approach would use periods such as days, hours, and even minutes for micro-batch partitions. We usually need to transfer tables between data warehouses or even to live databases. In this case, a typical query would be as follows for partition ranges, where *ds* is the date string, and *hh* is the hour.

```
SELECT * FROM dbdp.hourly_logs WHERE ds='2020-06-30' AND hh='13';
```

Partitions also make retention easier as we can drop partitions that we no longer need or archive them easily. Ideally, the job that loads the partitions should also do the retention to work as a compact unit.

7.2.3 Streaming

When we have dynamic data generation continuously, we might need to transfer the data directly to our analytics infrastructure. For instance, we might need to subscribe to one or more Kafka topics and write records to HDFS for later consumption. Streaming data is unbounded, as we have discussed. The caveat is about any aggregation that would be built on top of streaming transfers. The aggregation results would not match as more data would be accumulated in time. Nevertheless, the results might be accurate enough for many use cases. When aggregating streaming data sets with different data sets, we should be vigilant about the results and communicate them to the downstream consumers.

7.2.4 Timestamping

One of the biggest challenges is data dependency, which determines when it is safe to run the next batch of aggregation. When the tooling is the same, one can use the workflow orchestration to depend on other flows. Nevertheless, this approach sadly only works for small scale as there is no way to restrict different teams to use the same tool. Simultaneously, we still need to determine when it is safe to run an aggregation or mark an aggregation as complete. We can't rely on partitions for partitioned tables as we might get caught halfway through. The partition can exist but might not be full. Moreover, there is almost no way to check for unpartitioned tables.

The solution is to have a central authority for timestamping. The aforementioned system can be a file system, e.g. HDFS, where you have a directory for timestamps for each table in some format or service accessible by different clients. For HDFS, we can use file permissions to protect against writes to the directory. A simple HDFS-based directory stamping would be as follows: where we specify the data source, schema, and table name in the path along with the completed partition.

hdfs://dbdp/timestamps/dbdp_dw/dbdp/hourly_logs/ds=2020-06-30/hh=11.

HDFS can get a bit cumbersome since every client has to have an HDFS client setup. For timestamp service, we have to have groups or teams defined for particular tables to protect everything. Assuming timestamp service has a REST interface, we can do *GET* and *PUT* requests for timestamp checking and completion. The URL would look very close to the HDFS path as follows: *https://stamper/dbdp _dw/dbdp/hourly_logs/ds/2020-06-30/hh/11.*

7.2.5 Tools

There are many ways to transfer data from one data source to another. Cloud providers can have good integration within their platform. Thus, they can offer data transfer tools out of the box and even create a fully working data pipeline. Nevertheless, different teams might use different cloud providers, or the company can move from one cloud provider to another. Therefore, we will go through some useful tools that work in every environment, such as Sqoop, Embulk, and Spark.

7.2.5.1 Sqoop

Apache Sqoop is a tool to transfer data between Hadoop and other data sources. Sqoop relies on the database to gather table schema for the data to be imported or exported. Under the hoot, Sqoop uses MapReduce to import and export the data. Sqoop allows a couple of options. We can specify the table to import, and Sqoop will copy the entire table from source to target. Furthermore, Sqoop can accept a query that it can execute on the source and write results into the target. Nevertheless, the query option can be a bit adventurous if it is running against a production database. Sqoop can execute the copy operation with an option file to specify how to split data, a number of mappers, or delimiters. A sample sqoop option file would look as follows:

```
# Specifies the tool being invoked
import
--connect
jdbc:sqlserver://dbdp.mycompany.com:1433;databaseName=DBDP
--username
dbdp
--split-by
id
--num-mappers
3
--query
SELECT id, chapter FROM dbdp.chapters WHERE no_line > 2000
```

7.2.5.2 Embulk

Embulk is a parallel data loader tool that has gained recent traction and becoming a popular choice for data loading. Embulk supports transferring data between various databases, data storage, files in various formats, and cloud systems. Embulk can figure out data types from files, parallelize execution for Big Data sets, validate data, and recover from failures. Embulk uses a plugin-oriented approach to transfer data between data sources. Embulk has the following components:

- *Input*: The source of data. For example, PostgreSQL, Cassandra, HDFS.
- *Output*: The target for the data. For example, MySQL, ElasticSearch, Redis, CSV.

- *File parser*: The parser for different file formats. For example, JSON, Excel, Avro, XML.
- *File decoder*: Decodes files from different formats. For example, LZO, LZ4.
- *File formatter*: Formats files into different formats. For example, Avro, JSON.
- *Filter*: Filters out the rows from the table.
- *File encoder*: Compresses output to different formats. For example, LZ4, ZIP.
- *Executor*: Executes the transfer operation locally or on YARN.

An example Embulk configuration file to load a CSV file to PostgreSQL is as follows:

```
in:
  type: file
  path_prefix: /dbdp/embulk
  decoders:
  - {type: gzip}
  parser:
    charset: UTF-8
    type: csv
    skip_header_lines: 1
    columns:
    - {name: id, type: long}
    - {name: username, type: string}
    - {name: created_at, type: timestamp, format:
      '%Y-%m-%d %H:%M:%S' }
out:
  type: postgresql
  user: dbdp
  host: dbdp.book.server
  database: dbdp
  mode: insert
  table: users
```

7.2.5.3 Spark

Spark can come in handy as it can play a bridge role between different systems. The Resilient Distributed Dataset (RDD) abstraction and connectors make it easy enough to transfer data from one system to another. With the huge community support, Spark provides many opportunities to transfer data between systems. For example, we can easily read from Kafka and write to ElasticSearch or read from Cassandra and write to HDFS. Let us consider the following example: transfer from PostgreSQL to Redis in a couple of simple steps.

```
spark.read
  .jdbc(url, "member", props)
  .select("id" , "country")
  .write
```

```
.mode(SaveMode.Overwrite)
.format("org.apache.spark.sql.redis")
.option("table", "member_country")
.option("key.column", "id")
.save()
```

Spark transfers require a bit of programming effort, but it also offers additional transformation capabilities. We can create generic solutions between different systems depending on our needs and potentially create spark-submit wrappers. Once we have a library of wrappers, we can then easily transfer any tables between disparate systems. Additionally, we can easily set the number of partitions and parallelism to avoid exhausting either source or target system.

7.2.5.4 Apache Gobblin

Apache Gobblin is a distributed data integration framework that facilitates data ingestion, replication, and organization. Gobblin handles common data ingestion tasks from a variety of sources to sinks. Gobblin addresses scheduling, task partitioning, error handling, state management, data quality checking, and data publishing. Gobblin implements a framework that is auto-scalable, extensible, adaptable to model evolution, fault-tolerant, and deployable to multiple platforms (Apache Gobblin, 2020).

Working Modes Gobblin has different modes of working. We can import Gobblin as a library and run it as part of our infrastructure code. We can use Gobblin command-line interface (CLI) to execute one-off imports from one data source to another. Alternatively, we can integrate with our existing orchestration tools like Azkaban to launch Gobblin jobs. Moreover, we can have a standalone Gobblin service to run data transfer jobs continuously. To scale-out jobs, we can submit Gobblin jobs to YARN and leave the parallelization work to the YARN. These flexible options make it easy to adjust Gobblin jobs to different settings. Gobblin accepts job configuration files where we can list many properties about source and target. Below, you can see a job properties file where we transfer data from Kafka to HDFS.

```
job.name=DBDPKafkaToHDFS
job.group=GobblinKafka
job.description=DBDP job for Kafka to HDFS transfer
job.lock.enabled=false
kafka.brokers=localhost:9092
source.class=org.apache.gobblin.source.extractor.extract.kafka
  .KafkaSimpleSource
extract.namespace=org.apache.gobblin.extract.kafka
writer.builder.class=org.apache.gobblin.writer
  .SimpleDataWriterBuilder
writer.file.path.type=tablename
```

```
writer.destination.type=HDFS
writer.output.format=txt
data.publisher.type=org.apache.gobblin.publisher
  .BaseDataPublisher
mr.job.max.mappers=1
bootstrap.with.offset=earliest
```

Architecture Overview Gobblin offers flexibility and extensibility for different integrations. We can add new source/target adapters. To support ease of adaptation, Gobblin is built on top of constructs. A source represents the data source adapter and partitions data into WorkUnits. An extractor pulls the data specified by the source. Gobblin uses watermarks for the start (low watermark) and end (high watermark) of the data. Extractors hand over data to converters where they handle the transformation of the data on the fly. Converters are pluggable and chainable. After conversion, Gobblin uses quality checkers to validate the data. Once the data passes certain quality standards, it goes to writers where write to an output directory. Lastly, the publisher takes written data and commits it atomically (Figure 7.4).

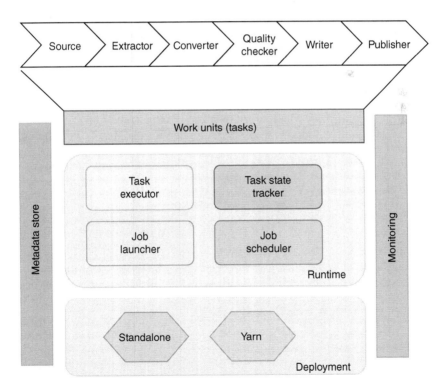

Figure 7.4 Gobblin architecture.

A Gobblin job has got tasks where each task corresponds to a work unit. A work unit is a portion of the data that needs to be retrieved from a source in a job run. Tasks get executed in the underlying runtime depending on the deployment setting. For development purposes, a standalone server might be a good choice, and YARN might be used for production purposes. Gobbling runtime handles job scheduling, monitoring, resource negotiation, state management, and task lifecycle.

In a typical Gobblin job run, the job gets scheduled and acquires a lock so that the same job does not run concurrently. The job gets the work units from the source and creates tasks. The job then runs the tasks, commits the data, and saves task/job states to durable storage. Lastly, it cleans up the temporary directories that the job creates during execution and releases the locks. When the job fails or gets canceled, it cleans up the temporary things it created and releases the lock.

7.3 Aggregating Big Data Sets

Data aggregation is the process of shaping the data into a meaningful summary. Once the data gets aggregated, we can query summarized data rather than going through raw data. Hence, aggregating data provides efficiency when data is queried. Data aggregation enables engineers, analysts, and managers to get insights about data quicker. Before aggregation, we have to move data into a common processing environment, as discussed in the first two sections. Once we move data, then we can cleanse it first, then transform it into the target summary and save summarized data with appropriate retention. Aggregation stages are depicted in Figure 7.5.

7.3.1 Data Cleansing

Data cleansing is the step where we can fix problems coming with data. We can remove incorrect, incomplete, or corrupted data, reformat when there is a formatting problem, and dedupe duplicates. Since we receive the data from multiple sources, each data source might bring its way of formatting or labeling data. If data is inaccurate or misleading, we cannot reach actionable and insightful results. Thus, we have to cleanse the data before processing it. The cleansing problem depends entirely on the data sources we depend. It varies drastically from one working set to another. Nevertheless, we will try to give some common scenarios to fix before starting the transformation.

Some of the data we receive might not be usable because the data might be partial, incorrect, and sometimes corrupted. We have to detect these problems and remove rows that don't conform to our data quality guidelines. The problem is

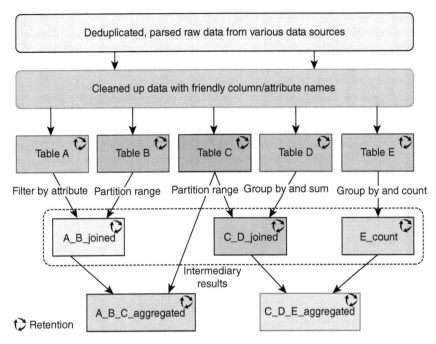

Figure 7.5 Data aggregation stages.

easier when some of the columns don't have data. We can easily filter. Nevertheless, it gets complicated when the data has incorrect data. For example, a country column might have correct countries along with some random text. We have to remove the rows with random text. An easy way is to execute an inner join with the country table. When the data is corrupted, we might not even load it properly to get eliminated when loading. Nevertheless, it is always good to have a threshold for loading. For instance, for every billion rows, 100 loading errors might be fine.

The next step is to fix the differences in data. We might get "True" as text from one data source and "1" from another one for a boolean value. We need to convert them into the same type. We need to go through all such columns and fix them before any transformation. Moreover, some data might come in completely different forms like JSON. We might need to explode JSON values into appropriate columns before joining with other tables.

The data might be almost complete but not fully. For example, we might have a city but not a country for some of the rows. We can potentially guess country value by observing the city name. Yet, we need to document and decide carefully about our observations. We never want to provide inaccurate data. It is probably better to miss the data than have fallacious data. The consumers of the data should not have doubts about accuracy.

Finally, some of the data might not align with the data trend, e.g. too extreme cases. We might want to smoothen data by removing such outliers. To detect them, we need to run some exploratory queries to find the top n elements for a given column or columns. Once we detect such outliers, we should come up with a reasonable threshold to remove them.

Consequently, we should review the data in terms of completeness, accuracy, and outliers. We should then develop a set of rules to cleanse data and add it as a step in our pipeline. We should also run quality checks on the data as well as running some anomaly detection.

7.3.2 Data Transformation

The transformation step executes a series of stages from one or more data sources into target data. Depending on the path from source to target, the transformation can be costly in time and effort. Yet, data transformation helps to organize data better for both humans and computers. We can produce an efficient data model for both querying and storage purposes. The transformation process utilizes transformation functions and stages that consist of common mutations such as filtering and aggregation. Transformations stages define the transition from one or more data sources to the target data.

7.3.2.1 Transformation Functions

Transformation functions let us mutate data from one more data source to the target. Some of the transformation functions change the data, but some of the transformations simply update the metadata. Metadata updates include casting, type conversion, localization, and renaming. Some common transformation functions that change data are as follows.

Filtering Data can come with irrelevant rows for what our target might need. In that case, we would like to filter these rows. For example, we might want to summarize API call-based data and remove any other HTTP calls.

Projection When summarizing data, we generally don't need all the columns. Thus, we would project data from a set of columns or attributes from the source data.

Mapping Mapping involves the conversion of one or more columns into a target column or columns. For example, we might combine two columns, weight, and height, to compute BMI (body mass index).

Joining The target data might need information from many sources, and we might need to join two data sources based on common keys. For example, we can join the customer action table derived from logs with the customer table.

Masking We might need to hide original data due to security concerns. We would hide the original data with some modifications.

Partitioning Most common Big Data systems have the notion of partitioning. We need to limit the amount of data we process by partitioning it into time ranges. A typical transformation might limit data into weeks, days, or hours.

Aggregation We can apply numerous aggregation functions over one or more data sources grouped by a set of columns. In some cases, we can even introduce our aggregation functions as well. Common aggregation functions are sum, max, min, median, count, and so forth.

7.3.2.2 Transformation Stages
Going from multiple sources to targets might involve a complex directed acyclic graph (DAG), where we would have to output intermediary results. We might use these intermediary results in other flows. Most commonly, we need these results simply because the process would become cumbersome otherwise. There are a few areas we need to harden for the transformation stages.

Timestamping We already discussed timestamping in data transfer. We should timestamp the data for intermediary stages as well. Flows outside of the current one might depend on the intermediary stage to complete their transformation. Timestamping will ensure that the data won't be incomplete for other flows.

Validation Important steps in complex transformations should have relevant validations and anomaly detection. Validations and anomalies should fail the entire transformation. We don't want to waste resources on computing inaccurate data. More importantly, we shouldn't produce misleading data.

Idempotency Retrying any step in transformation should always give the same effect. The steps in transformation should always clean up before rerunning the same step. For example, we should clear the directory for the data before executing the transformation step.

Check Pointing When a single transformation step gets complex or takes a long time to compute, we should consider it to divide it into multiple stages. Multiple stages will allow us to fall back to the previous stage when a stage fails. Furthermore, multiple steps allow optimizing the transformation easier.

7.3.3 Data Retention

We should consider data retention as part of the aggregation process. We should decide the retention policy for each aggregation step. Having well-defined retention on steps offers advantages for downstream data consumers and system admins. Consumers can know how long the data is kept to set up their business logic accordingly. System admins may not need to execute manual or semiautomated processes to free unused data. We should also have a retention policy for intermediate stages of data. Having a retention policy on intermediate results can help us in reacting to bugs and data problems easier. We can run transformation steps with appropriate retention where the bug was introduced and skip the previous ones.

7.3.4 Data Reconciliation

Data reconciliation is a postprocessing method to improve the accuracy of the data by handling missing records and incorrect values. Reconciliation jobs work with updated data sources to produce the most accurate results. Ideally, all of the data come cleansed to the aggregation layer. Nonetheless, data might deviate due to integration failures, upstream data format change, and many other factors. In some cases, reconciliation might need to coexist with normal aggregation because we deal with a stream of data. For example, if we are processing data daily, we might need to cut out a user session in the middle of the night. We can correct this behavior with a reconciliation job that finds out such cases and updates them.

Although we must aim for perfect accuracy, it is not possible. When it comes to reconciliation jobs, we must define an error margin to trigger them. If the error rate is under the error margin, we should skip postprocessing. If the error margin is too high, we might serve unexpectedly inaccurate data. However, if it is too low, we would waste resources for unnecessary postprocessing. We should find a nice balance for error margin depending on the business requirements.

When we are running reconciliation as part of our data pipelines, we should avoid rewriting the whole partitions if possible. One way to deal with such operations is to writing reconciliation updates to a partitioned table with the same partitions as the original aggregation. When reading for accuracy, we can consolidate results from both tables and return the most accurate answer to the user. In many cases, a view over both tables might serve users very effectively, e.g. avoiding the whole consolidation.

7.4 Data Pipeline Scheduler

One of the central components of a modern Big Data platform is workflow management. The Big Data platform should offer a workflow management solution that makes it easy to set up pipelines and reliably execute them. Although there is no comprehensive solution to solve most of the day-to-day data operations, we still have systems such as Jenkins and Airflow to cover most of our needs. In an ideal world, a workflow management tool should address many different needs as follows:

- *Scheduling*: The tool should trigger pipelines on complex schedules.
- *Dependency management*: The tool should make it visible which task depends on which one.
- *Workflow*: The tool should offer a pipeline as a unit and provide context between tasks it runs.
- *Credentials*: The tool should support different credential management policies.
- *Execution*: The tool should execute tasks in a horizontally scalable fashion with retries on individual tasks.
- *Communication*: The tool should integrate with common communication channels such as e-mails, notifications.
- *Local testing*: The tool should have local validations as well as running on test data before publishing a pipeline to production.
- *Visualization*: The tool should visualize workflows, tasks, and metrics.
- *Resource management*: The tool should be aware of both internal and external resources. Queuing and priority management should be an integral part of task execution.
- *Ease of use*: The tool should provide easy ways to create a pipeline through the user interface. This is more of an advanced feature, but it would make things so much easier for non-developers.
- *Ownership*: The tool should have a sense of group ownership for pipelines as well as artifacts, credentials.

Notwithstanding, workflow tools have some of these features but not all. Depending on our needs, we need to choose the right tool for the Big Data platform. We would like to go over some of the common workflow management tools widely used for Big Data pipelines.

7.4.1 Jenkins

Jenkins is a well-known automation tool that offers various features that can be used in scheduling data pipelines. We can do advanced job scheduling to run a Jenkins pipeline with stages as steps in the workflow. Jenkins logs everything quite nicely, alerts on failures, and has an amazing number of plugins we can use. Moreover, we can install various client software such as hive, PostgreSQL (psql), spark, and more for Jenkins. With the help of client tools, we can then run our data transfers and aggregations on Jenkins. Nevertheless, Jenkins lacks the visualization for the workflows, and dependency management can get tricky for larger infrastructure undertakings.

Using Jenkins pipelines, we can create a multistage workflow where we call different scripts for a different set of operations. The scripts can set up the environment and run the required job. In the following example, we copy the table from PostgreSQL to HDFS. Later, we aggregate the member table with logs. Note that the example has certain assumptions, e.g. credentials. It would take a bit of time to set such a pipeline initially.

```
pipeline {
  agent any
  stages {
    stage('copy-member-table') {
      steps {
        git 'https://git.dbdp-examples.com/dbdp/dbdp-jenkins.git'
        withCredentials([string(credentialsId: 'dbdp-psql',
        variable: 'PW')]) {
          sh "./bin/copy_from_postgesql.sh public member dbdp ${env.PW}"
        }
      }
    }
    stage('aggregate-member-with-log') {
      steps {
        sh "./bin/aggregate.sh ./sqls/aggregate_member_with_log.sql"
      }
    }
  }
  post {
    failure {
      mail to: 'dbdp-dl@dbdp-examples.com',
        subject: "Failed Data Pipeline: ${currentBuild.fullDisplayName}",
        body: "Something is wrong with ${env.BUILD_URL}"
    }
  }
}
```

7.4.2 Azkaban

Azkaban is a distributed workflow orchestration system. Azkaban allows running flows sequentially by specifying dependencies. Azkaban offers rich workflow visualizations, job summaries, statistics and flow management, complex schedules, and more. Azkaban supports Hadoop, Hive, Pig through job types. Nevertheless, Azkaban falls short when it comes to integrations with data processing tools. Hence, we have to install most of the tooling on our own. Azkaban has a base job type called command, where we can execute any shell script. As we have suggested for Jenkins, we can create a library of wrapper scripts that can execute a different set of operations such as data transfer, aggregations, calling an external service. By using command type, we can then execute any of the scripts from the library. Fortunately, Azkaban offers rich user interface features that can make our life easier when coping with multi-flow pipelines.

7.4.2.1 Projects

Azkaban works with projects. A project can have one or more flows. Each flows consists of jobs. Jobs can be executed in parallel when possible if the number of executors is set to more than one. Each job consists of a job type with related configuration and dependencies. A job looks as follows:

```
description=imports a completed aggregation to the data mart
type=command
command=./import.sh dbdp click_conversions_by_member
dependencies=click_conversions_by_member
```

Azkaban jobs assume that all jobs are synchronous; therefore, they can't wait on external resources asynchronously. Every dependency outside of Azkaban needs to be handled through busy waiting where we check periodically if the dependency is satisfied or not. If everything were in Azkaban, this model wouldn't have been a big problem. Nonetheless, that is mostly not the case. Moreover, suppose the amount of jobs that have to do busy waiting goes up. In that case, it is highly plausible to experience performance problems as the system is wasting resources for the sake of busy waiting. Another downside is searching. It is not possible to search flows in the project. One has to know which project contains a specific flow. Otherwise, the user interface doesn't offer much about searching.

7.4.2.2 Execution Modes

Azkaban can be executed in two modes. The executor and web server runs inside the same process with the H2 (in-memory) database in solo mode. This might be enough for simple use cases, but we would not like to lose data or depend on such settings when executing mission-critical data pipelines. In multiple executor mode, web servers and executors are separate. A MySQL database backs them by active/passive configuration. Ideally, the web server and executors are in different hosts, so problems on the web would not cause executors or vice versa.

In a multi-executor setup, we have to specify each executor through an IP address in the database. In case there is a problem with one of the executors, we cannot replace it automatically. We have to intervene manually. Another downside for executors is monitoring. Azkaban exposes JMX metrics with a little bit of documentation. Nevertheless, Azkaban is pretty stable and does not cause many problems when running in production when bare metal instances are used.

7.4.3 Airflow

Airflow is a workflow automation tool to programmatically author, schedule, and monitor workflows. Airflow uses DAGs to represent a workflow containing all the tasks we want to run in an organized way. A DAG is a python program where we define operators and the relationship between operators programmatically. An operator is a step in the workflow. Each operator becomes a task in the workflow. An operator usually should be atomic such that it does not carry information from previous operators. Airflow comes with python, bash, e-mail, and many database operators out of the box. Airflow also has a special operator called Sensor that waits for an external dependency (Airflow, 2020). Let us now see an example airflow DAG.

```python
import os
from datetime import datetime

from airflow import DAG
from airflow.operators.bash_operator import BashOperator

DIR = os.path.dirname(os.path.realpath(__file__))
SQL_PATH = os.path.join(DIR, 'sqls/customer_summary.sql')
GIT_PULL = os.path.join(DIR, 'bin/git_pull.sh ')
EMBULK_IMPORT = os.path.join(DIR, 'bin/embulk_import.sh {} ')
RUN_HIVE_AGG = os.path.join(DIR, 'bin/run_hive_agg.sh {} ')
tables = ['customer', 'customer_detail', 'customer_action', 'customer_order']

# we automatically assign new operators to the dag with context manager
with DAG('dbdp_dag',
    description='Import from various databases and aggregate',
    schedule_interval='*/5 11 * * *',
    start_date=datetime(2020, 7, 2), catchup=False) as dag:

    update_sqls = BashOperator(task_id='update_sqls',
        bash_command=GIT_PULL)
    customer_summary = BashOperator(task_id='customer_summary',
                                    bash_command=RUN_HIVE_AGG
                                    .format(SQL_PATH))

    for table in tables:
        table_import = BashOperator(task_id='import_{}'.format(table),
                                    bash_command=EMBULK_IMPORT.format(table))
        # shift operator is used to chain operations
        update_sqls >> table_import >> customer_summary
```

In the example DAG, we use a couple of scripts to import tables from the customer database to aggregate them with processed log data. We have three wrapper scripts to get recent changes, import tables with Embulk, and run a hive aggregate function. Note that each of these operations can be a specialized operator like Hive operator. Nevertheless, we rather have chosen to wrap the commands into scripts.

7.4.3.1 Task Execution
When the above DAG runs, airflow converts it into a DAG run, an instance of the DAG. It instantiates operators as tasks and each run of a task as a task instance. A task goes into different stages from start to finish. In a typical run, a task goes into

scheduled, queued, running, and success. If the task fails and if it can be retried, it starts back from scheduled. Airflow provides additional functionality like pooling and queuing.

Pools provide resource management mechanisms. Pools limit the execution of parallelism on the resource. For example, we can define a pool for MySQL, constraint the number of tasks Airflow executes by setting the pool for MySQL tasks. Controlling resources also necessitates assigning priorities to tasks. Thus, Airflow allows setting priorities to tasks. The executor performs tasks based on their priority. The pooling is a nice approach. However, Airflow does not support multiple pools. A task might depend on multiple resources, and each resource might have constraints. In this case, the system should evaluate resources on multiple pools to let a task run. Nevertheless, supporting multi-pooling makes resource management even harder since the system has to meet more requirements.

Moreover, we can also set queues for tasks. By default, every task goes into the default queue. We can define queues and set workers to consume tasks from one or more queues. Queues are useful when some tasks need to run on a specific environment or a specific resource. For example, we can set a queue for Hadoop tasks and have a working environment for Hadoop tasks. Whenever a Hadoop task comes in, it goes to Hadoop ready environment.

7.4.3.2 Scheduling
Airflow monitors all the DAGs in the DAG folder. It finds out tasks that have dependencies met and schedules them to run. Periodically, it parses all the DAGs and checks out dependencies for active tasks to see if they can be triggered. Each DAG can be assigned a schedule. A schedule can be any cron argument, where we can run our DAG daily, hourly, and more. Additionally, Airflow supports catchup and backfilling.

A DAG can define a start date, end date, and a scheduled interval. Between the start date and end date, intervals define how many times a DAG can run. Airflow will examine the intervals that the DAG is missing, and it will kick off any missing interval in the past until it catches up. If the DAG knows how to catchup itself, then catchup can be disabled.

Airflow can backfill the previous intervals by the backfill command. A backfilling process is rerunning previous intervals by a specified date range. The backfill command would prompt the user if they want to clear previous runs. Airflow will then run intervals between the specified backfill range automatically.

7.4.3.3 Executor
Executors provide the mechanism to run task instances. Executors in Airflow are configurable. Depending on the requirements, we can choose an executor out of available executor options. Airflow comes with a couple of executor options. We will briefly look at Celery and Kubernetes executor.

Celery Executor   Celery executor uses celery, a distributed task queue, to distribute work across multiple workers. To enable the celery executor, we have to install additional dependencies, but it is rather straightforward. Underneath celery, we can use RabbitMQ or Redis as a message broker. In the celery executor setup, the scheduler puts the commands to be executed to the celery. Workers get commands from celery and execute them. Workers then save the task status back to the celery. Scheduler gets status information from celery about completed tasks. Web server fetches job logs from each worker. Note that each component in the system needs to access DAG files, so it is the operator's responsibility. A simple cron job that periodically pulls the repo should be more than enough for synchronization for most of the cases. The celery executor architecture can be seen in Figure 7.6.

Kubernetes Executor   Kubernetes executor enables Airflow to scale very easily as the tasks run on Kubernetes. There are a couple of integration points that need to be addressed, like DAG files or logs. We can use Kubernetes persistent volumes for DAG files and read them as usual with the same setup. For logs, we can again store them in the persistent disk. Another option is to store them in an object store and stream them from the object store when requested.

DAG files introduce consistency problems since they are refreshed constantly, and they are not tied to the execution. A way to overcome this problem might be to integrate with git by default, and use git commits to track which version of DAG the system is executing against.

7.4.3.4 Security and Monitoring

Airflow supports all well-known authentication methods and allows them to integrate with others. Airflow has user permissions but does not support group-level permissions. Group-level permissions make it easy for teams to own workflows. Additionally, Airflow supports securing API and connections.

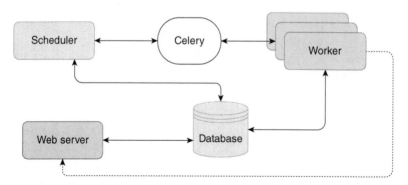

Figure 7.6   Celery executor architecture.

Airflow integrates with statsD, a daemon for stats aggregation and with sentry, an application monitoring and error tracking tool. It has counters, gauges, and timers for a different set of metrics.

7.4.4 Cloud

Cloud providers offer automation tools to copy log files, execute them on an on-demand cluster, and write results back to their warehouse offering. Most of the actions can be done through the web user interface or the CLI. Thus, onboarding is easy and efficient. The security aspect of the steps is easier since we are within the same cloud provider. The downside of using cloud pipelines is the dependency on the cloud provider. It is very costly to migrate from one provider to another since the whole workflow is tightly coupled with the provider-based tooling.

7.5 Patterns and Practices

Building effective Big Data pipelines require assessing the relative merits of data in the context of business. The data pipelines address the need and requirements of the data analytics process as a whole. When designing and implementing data pipelines to deliver business value, we must consider several aspects. This section would like to go over some of the data patterns, anti-patterns, and best practices for data analytics.

7.5.1 Patterns

Employing effective patterns can help in every part of software development. Data analytics is no different. Thus, we will go over data centralization, a single source of truth (SSOT), and data mesh patterns. Nevertheless, Big Data analytics is still evolving fast. These patterns can change drastically over time. On the contrary, these patterns are widely adopted in the industry. I also believe utilizing these patterns helps in many ways.

7.5.1.1 Data Centralization

When data lives in multiple environments without a chance to bring them together, figuring out answers for simple questions can be unpleasant. Having data in separate systems is a common problem. Companies have some customer data in customer relationship management (CRM) systems, service data in logs, and advertisement data in a service provider. Without bringing all this data into common data storage, it is tough to get a broader customer view. That is one of the reasons why data warehouses and data lakes became widespread in the industry.

Centralizing data can lower the cost of storage. The common storage can be a lot cheaper than other options. We can then optimize the data, which has long-term value. What is more, having all data in one place can increase the consistency among the data processing pipelines. Last but not least, centralizing can also bring more opportunities between different sets of data without any cost. Aggregating data becomes an easy step rather than a convoluted process.

In consequence, an extract, transform, and load (ETL) pipeline starts with transferring necessary data from multiple data sources to a common storage layer. The transfers can happen between a relational database and a data lake or service provider API and data lake. The Big Data platform should be able to support any type of transfer in a scalable manner.

7.5.1.2 Singe Source of Truth

SSOT is prevalent in data management. In short, SSOT represents a data set that the organization agrees upon and trusts its quality. The need for SSOT comes from the scattered nature of Big Data in different systems. When the data is centralized, it still has to be processed. Nevertheless, organizations tend to implement the same type of data pipelines in separate pillars. Implementing similar solutions cause inefficiencies, inconsistencies, misinterpretations, and deficiencies.

Having a different team of people working on the same type of pipelines results in a loss of resources. The teams can come up with models that refer to the same data but are worded differently. Hence, there would be discrepancies between the two data models. Between these two data models, interpretations can differ uncontrollably. An analyst looking at one model expects the same from the other one. In reality, there could be many subtle differences. In the long run, additions and updates to the data models would make disparities even worse. The data itself will be under investigation, but any conclusion from these models would pose a risk to the business. Hence, implementing SSOT becomes an excellent way of dealing with such strains.

Naturally, awareness of problems within data and its schema happens with time. At some point, it becomes quite hard to onboard someone to existing data models. Thus, the implementation of SSOT becomes almost a necessity rather than a choice. The best way to approach those problems is to consolidate models thoroughly with the information we have. Later, we should identify common needs from the business and group them into a set of data sets with corresponding pipelines. Each of these data sets is owned by the organization as a whole. Any problems in the data set or generating pipelines would escalate to each of the business pillars. Changes in these data set should be reviewed from different parts of the organization so that the organization can keep track of institutional knowledge and the change log.

SSOT requires data centralization. All the relevant information has to be brought into the same data storage. Once the data has been transferred, we can set up filters, cleansing, and aggregations to deliver SSOT tables. Resulting tables can be transferred to data warehouses, data lakes, or even live site databases for decision-making purposes. Since everyone knows the meaning of data and sure about the quality, different parts of the business can react faster to new business needs by leveraging SSOT.

7.5.1.3 Domain Driven Data Sets

The domain-driven-design focuses on business entities. Our approach for domain-driven data sets is similar. We also concentrate on business entities for Big Data sets. We ingest data into the platform from various sources. In all of the sources, we have a schema that represents a domain entity. When we parse the logs, they also represent a business entity, application access.

All in all, we get various domain entities into our storage system, where we process the data for further consumption. The analytics processing also involves business entities and domain since we want to deliver business value. One approach is to represent everything in a single superset. Nevertheless, this is not attainable as the business has branches and has specific requirements. Thus, grouping data based on the domain to data sets makes it easy to maintain and relate to the business purpose.

Coupling irrelevant business entities create unnecessary complications. Cleansing, filtering, or further aggregation becomes more laborious since pipelines have to process all in one place. Each coupled business entity will bring its business rules. Nevertheless, the business rules are only related to one domain. Moreover, the operational cost will increase drastically, as multiple business units have to be coordinated to change. Besides, it is hard to tell ownership of the data.

The alternative is to define data sets based on the domain. Some of the data for all domains can come from a base domain like application access. Nevertheless, the rest of the data should be divided into domain-driven data sets. When one or more domains need to be merged for cross-functional insights, merged data will represent another domain. The key here is to define the domain and related entities. The domain should be related to the business value, and a team who knows the business purpose should support it. Therefore, business-related rules can be applied to the domain data set, and the team can be responsible for it. I would like to depict an example assignment of domain-driven data sets as in Figure 7.7.

In the example, we have pipelines that cleanse access logs, filter them, and aggregate them differently. The application access pipeline serves the application access domain that other domains mostly depend on. Domain-driven pipelines create their data sets with the domain they depend on and additional data brought from the data layer. When two domains merge into a new one, the data models from

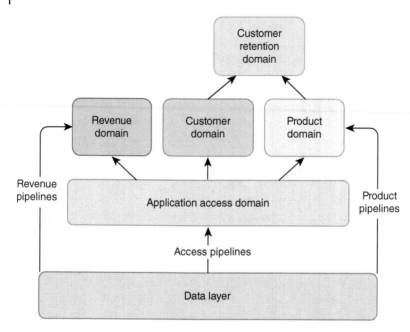

Figure 7.7 Domain-driven data sets.

both domains are used to generate the new domain data set. Each domain-driven data set represents a business unit or function and is strongly integrated with the business purpose.

7.5.2 Anti-Patterns

While trying to solve Big Data analytics problems, we might end up in anti-patterns solutions. At first glance, the solution might not look like an anti-pattern; however, it can turn into an anti-pattern with time. There are two anti-patterns we have observed while designing data pipelines: data monolith and data swamp.

7.5.2.1 Data Monolith

Monolith has become an infamous word in the world of microservices. Nevertheless, the monolith represents an important reality. Our product or service has become a huge success. It has grown so big that it is hard to cope with it in one place. The solution has been smaller components such that responsibilities are delegated to the smaller systems. The approach aligns with the single responsibility principle in a wider context.

Data monoliths are no different from application monoliths. We started collecting information from one or more places and added more columns to our denormalized models. At the time, it made sense to do so operationally and organizationally. Nevertheless, the models got bigger and bigger. The context on the columns got lost. The business has changed into directions such that some of the columns aren't even relevant anymore. People who worked in different parts of the monolith aren't with the company anymore. There is no reliable source of information about the columns.

The solution to the data monolith resembles a solution to an application monolith. We start with the information available on the data monolith and identify the responsibilities. Later, we start moving out of these responsibilities to their separate pipelines and backfill the data with older data from the data monolith. We repeat the cycle up until we don't have any dependency on the monolith.

For future data sets, we closely maintain domain boundaries on the models to grow out of their business context. Once we establish the routine for the domain-driven data sets, we can avoid future monoliths.

7.5.2.2 Data Swamp

Data lakes provide the opportunity to store structured, semistructured, and unstructured data for data analytics. Data lakes make it easy to gain cross-domain insights. Pouring all data into the data lake has been well-expected behavior. If the data gets chaotic without an active management strategy, the data lake becomes a data swamp. A data swamp lacks data governance, data lifecycle policies like retention, organization of data sets. Applying data analytics practices in a data swamp becomes a nightmare. There is no trustworthy information, and every piece of data is scattered around. Standard ETL operations take longer since metadata is missing, and most of the data needs parsing and deep cleansing.

An organization can end up in a data swamp when there is no established policy for the data lake. There should be organization-wide policies on how the data will be poured to avoid such problems. Moreover, data ownership needs to be employed regardless of the data structure. Each ETL pipeline should have well-defined sources and with a team owning it. Ideally, pipelines should bring most of the data into the system with a schema representing a business model. Schema on-read has its advantages, but it shouldn't end up being standard practice. However, having too strict rules might also hinder various parties to experiment with different data sets. Thus, creating balanced policies for data management and governance for the organization is to deal with such a problem. The policies might differ from one organization to another.

7.5.2.3 Technology Pollution

Thanks to many open source projects, the vast amount of tools we can use for Big Data platform is impressive. We get more and more tooling for solving similar problems. Although staying up to date with technology is a good move for the long term, establishing policies on tools is also important. Having a new tool in the system would mean operational burden and the need to support, maintain, and upgrade a new system. Even if we can outsource operational costs, we should still be careful since there is always a certain learning curve for new systems. Having so many technologies for analytics processing would create turmoil in the data pipelines.

When establishing data pipelines, it is important to choose standard technologies for data transfer, consolidation, aggregation, and scheduling. Ideally, most of the tooling is decided companywide, so it is not so hard to transfer knowledge between teams. Onboarding a new employee to the system wouldn't require so much time as the tooling stabilized. Furthermore, coping with Big Data analytics tools isn't easy. It takes time to get used to some of the pitfalls of different tools. Having experience in a certain tool and transferring this knowledge would help people experiencing the same issues.

7.5.3 Best Practices

Applying patterns and avoiding anti-patterns can help institutions in the long run. The patterns and anti-patterns can depend on the context. Thus, we would like to further strengthen our approach by applying some of the best practices.

7.5.3.1 Business-Driven Approach

Starting Big Data analytics with a shiny toolset has been the case for many companies. Nevertheless, it is hard to have a direction without a well-defined business goal. Before investing in both infrastructure and resources, we should set a business goal. For example, a goal can be delivering partner reporting on products they sell on our platform daily. With this clear goal, we can then reason about where to collect data, store it, and process it. All of the options are up to evaluation. It depends on what we have so far and how much effort we can put into this particular problem.

Occasionally, we might have to undertake large infrastructure projects depending on the requirements. For instance, we might have to change how we do data analytics, such as changing our pipeline tooling. Even for such infrastructure projects, we have to have a clear business purpose. It doesn't have to be directly related to the business. It can be related to developer productivity, cost reduction, and so forth. With this in mind, we can justify the project timeline and show the progress we made in quantifiable results.

7.5.3.2 Cost of Maintenance

Maintenance costs are often disregarded when there is so much optimism about technology or tooling to use in our analytics pipelines. Every new tool, every new technology comes with its maintenance cost. Even if we can outsource part of our infrastructure to a cloud service provider, we still have to put in some operational effort. Maintenance costs are ongoing costs. We can lower them a bit with some automation. Yet, we still have to include them in key decisions for our data analytics processing.

If we want to leverage a new technology or tool, we should consider what it would cost to run this system. How much training do we have to go through to support the system in production? There is a significant difference between using a system for development vs. production. Sadly, many problems occur only in production environment. What is more, sunsetting a production system takes a huge amount of effort. When we bring the new sparkling tooling, we should also consider our exit strategy. This is even more important when we outsource our infrastructure to a cloud provider.

Consequently, any new update to the data analytics infrastructure should go through a thorough review of maintenance costs. Ideally, we should have a rough idea of how many people we need to support the system, how many machines it takes, how we scale it on demand, and so on. Although we recommend carefully evaluating maintenance costs, it should not also make a road blocker. We can always start small, maintain less, measure demand, and put efforts accordingly.

7.5.3.3 Avoiding Modeling Mistakes

Data modeling is an important step in delivering key insights about the business. A strong data model can deliver highlights of the business clear, understandable, and consumable way. There are so many ways the data get poured into the data storage. Cleaning this data and aligning the model with the business is a big challenge. The model should be both clear and understandable so that the adoption would be quick. The data model would fall short when it does not meet business requirements, lacks domain context, and hard to consume.

Often, engineers focus on convenience rather than effectiveness in modeling. The primary goal of any data model is to satisfy a business requirement. When the model becomes convenient for development but not how the business needs the model, it loses its value proposition. Hence, when modeling the data, we should optimize for business. How we get there is a detail for us to figure out. Suppose we can understand the business purpose, long term, and data and how the business may evolve it. In that case, we are better positioned to model the data better and facilitate analytics faster.

The domain drives the model. If we fail to grasp all the details for the domain, any model we would produce will lack effectiveness. There are subtle details in every domain, which makes a huge difference in the model. There may be decisions in the past for the domain and may not seem the most optimal now, but were dictated by opportunity cost. We have to capture these details and include them in the modeling of the data.

Lastly, the model we produce should be easy to digest. When the domain is complex, and there are so many business requirements for the model, it might be hard. Often, engineers come up with complete solutions, and there is not much to do at this point. We either accept the solution and move on or reject and trash the engineering time spent on it. Even if the model satisfies requirements, it might be very hard to understand and takes a lot of resources to produce it in data pipelines. Thus, it is better to ask for feedback early on from potential consumers of the data. With a couple of iterations on the data model, we can deliver an effective version.

7.5.3.4 Choosing Right Tool for The Job

The first and foremost important thing for choosing analytics tools is starting from what we already have. Our data analytics toolset might not cover all the cases, but it might be good enough for what we are about to process. There are two benefits to choose a tool from existing tools. The first one is maintenance. We don't need to introduce another tool, evaluate it, and maintain it in our stack. The second reason is developer productivity. A well-known tool within the company would make the adoption and maintenance of the data analytics pipeline easier. There is always some extra friction when a new technology or tool has been employed since it takes time to adjust and learn these tools.

When introducing new tools for data analytics pipelines, we have to get engineering units to buy-in to the tool. As we mentioned, using a tool in development vs. production is a completely different experience. The safest approach is to introduce the new tool in phases. In the first phase, the tool can be in an alpha mode where there is not much promise about stability. Once the tool gets adopted and solves some important business problems, then it becomes a viable option to invest additional resources to use it in our analytics toolset. Big Data analytics is an evolving area; therefore, we can encounter many problems regardless of the tools we choose. Some caution is necessary, but it should not block us from experimenting with new tools.

7.5.4 Detecting Anomalies

Maintaining data quality is one of the notable challenges of data analytics. With so many moving parts in ETL pipelines, we can often end up with surprises. We can deliver an incorrect report to upper management, share inaccurate data with partners, and create troubles in production systems. Delivering nothing or delivering late is better than delivering misleading data. We have to build a mechanism to prevent us from delivering incorrect data. This is where anomaly detection kicks in. We believe every pipeline that generates data sets should have a couple of anomaly detection checks to notify data owners and prevent delivering key tables to the interested parties.

I define a data anomaly as an event or observation that does not conform to an expected computed pattern. An anomaly event occurs when there is a significant difference between an observed value and an expected value. Anomaly events can be detected in two ways. The first one is manual anomaly detection, where we write verification queries. The second one automated anomaly detection, where we can build rules to run on column variations to verify data consistency.

7.5.4.1 Manual Anomaly Detection

Manual anomaly detection involves simple boolean checks. We fire off an anomaly notification if a certain observed value is not within the expected range. These types of checks require a good grasp of domain knowledge and business expectations. We need to have a good idea about what an anomaly behavior is and how to write a query to filter out such events. We can observe anomalies by looking at various aggregations such as sum, max, min. For instance, we can check that an item price per transaction can't be higher than 1000.

When we write anomaly queries, we must have a clear expectation. We should write these checks to take many factors into account, like seasonality, time, etc. For instance, we might want to check the number of video game shoppers over the last few days. If the number is quite different from what we expect compared to the monthly or weekly minimum value, we can think that there might be an anomaly. Perhaps, new application release does not show up video games in the right category. We would like to fire off an event even if it is a false positive. We don't want to miss critical problems, so it is better to be conservative. It would take some extra querying for follow-ups, but catching the problem is an important part.

7.5.4.2 Automated Anomaly Detection

Automated anomaly detection involves a comparison of columns to the recent values observed. If the difference is too big, then we can say there is an anomaly. The logic for automated anomaly detection can get very complicated. We can write a framework to receive a table with columns and make some statistical analysis on the table with the given columns. Nevertheless, we can do simpler checks like table row count or a number of null values in a column.

At the very least, we should have automated checks like row counts. This will prevent disaster scenarios. Let us imagine the following scenario. We have two tables. We subtract one table from the other one. If the table we are subtracting is not populated, then the result would be all of the rows from the first table. We don't want this result. If we had a simple anomaly detection where we checked the counts, we could identify the problem before delivering the result and halt the processing.

More sophisticated anomaly detection can mean statistical analysis of columns and even machine learning models to detect unexpected results. Indeed, such a framework requires resources from the company to build. If we have such a framework, we can include it in critical sections of our data pipelines.

7.6 Exploring Data Visually

Interactive data visualization connects the dots with analytics. Visualization solutions provide business intelligence capabilities like dashboards and reporting. They are great for gaining insights about different dimensions of data without worrying about querying. There are numerous great solutions in the data visualization area. Some of them are paid products, and some of them are open-source products with enterprise versions. We will briefly touch base on open-source products, Metabase, and Apache Superset.

7.6.1 Metabase

Metabase offers some neat solutions for business intelligence. Once we setup up our backend databases like Presto or SparkSQL, we can start exploring our data. We can navigate to ask questions or browse data, which gives a quick exploration of our tables. Metabase provides some quick facts about tables such as distribution, multiple counts, and so forth. When we dig deeper into an individual column with X-rays, Metabase computes counts and distribution for the column. Metabase comes with dashboards and pulses. Dashboards are a collection of charts that we are interested in. Pulses are a notification mechanism for the questions in the dashboards.

A dashboard has multiple questions where a question is a dimension of data we want to know. Dashboards generally have a theme such as orders, logins. An order dashboard may display the number of orders for the last week, orders by price, and so forth. We can organize dashboards such that they show key performance indicators (KPI). We can then share them with our team or executives to see how the company is doing concerning the KPIs we have defined. Metabase allows zooming into the individual questions and getting more details for that particular question.

If we want to send our findings to partners or other interested parties, we can do so with Pulses. A Pulse is a notification mechanism in Metabase. A typical notification mechanism is an e-mail. We can easily set up our e-mail and send notifications.

7.6.2 Apache Superset

Apache Superset is another open-source business intelligence software that provides a pleasant dashboarding and chart experience. Apache Superset is designed to be highly available and scalable. At the time of writing this book, it is an Apache incubator with a strong community. Superset integrates with a wide range of database systems.

Superset brings many rich visualization types. We can explore, filter, aggregate, slice, and dice our data visually. We can create an interactive dashboard based on the charts we create and share them with partners, teammates, and the executive team. Moreover, it comes with SQL Lab, where we can run SQL queries against the connected databases.

Apache Superset can scale horizontally pretty easily behind a load balancer. We can run as many superset containers as we need for the front end and workers. It uses Celery to distribute querying tasks across workers through Redis. Superset further leverages Redis or Memcached for chart caching.

8

Data Science

After reading this chapter, you should be able to

- Describe data science applications.
- Understand the life cycle of data science.
- Use various data science tools.
- Use tooling for data science production.

Companies want to use any data for competitive advantage. Manual analysis of big data is difficult and tedious. An alternative might be applying data science techniques to analyze the data automatically. Data science applications' Big Data might provide customer satisfaction and retention gains with the data science tools and algorithms on the rise. Therefore, a modern Big Data platform should embrace data science from the bottom up.

The data science process includes several techniques. From data to actionable results, one might employ a mix of strategies. Some strategies might involve business understanding, domain knowledge, and creativity. Some strategies are relatively straightforward as it is left to the underlying algorithm or tooling like pattern discovery. Although there have been significant data science developments, the number of easily applicable techniques is arguably stable. This chapter discusses data science techniques, applies them through various technologies, and discusses deploying data science projects on production.

8.1   Data Science Applications

Data science has been a vital technology for organizations to decide, predict, and discover. Industries ranging from finance, manufacturing, transport, e-commerce, education, travel, healthcare, and banking have been employing data science

Designing Big Data Platforms: How to Use, Deploy, and Maintain Big Data Systems,
First Edition. Yusuf Aytas.
© 2021 John Wiley & Sons, Inc. Published 2021 by John Wiley & Sons, Inc.

applications for different problems. Banks use data science for fraud detection, manufacturing for problem prediction, and healthcare for the diagnosis. Although the problem space for data science is quite large, we reduced it to three categories: recommendation, predictive analytics, and pattern discovery. However, there are data science applications that might not fall into these categories.

8.1.1 Recommendation

Recommendation offers benefits to many organizations in different domains. Each organization has a unique set of recommendation problems, and each recommendation problem has its own set of goals, constraints, and expectations. Data science can help to find recommendations based on the similarity or closeness of items or individuals. Questions like which people the user might want to connect, which songs would user might like, or which products the user might want next can have answers.

Link prediction is one of the techniques to find links between items or individuals. It tries to find new connections between items or individuals. Link prediction algorithms can estimate the strength of a link to recommend higher scored links first. Link prediction techniques can guide new social connections to a user. Moreover, recommendation systems can personalize user experience through collaborative filtering, content-based recommendation, and hybrid approaches. Collaborative filtering works by building a data set of preferences for items by users. A new user gets matched against the data set to discover neighbors, which are other users who have historically had a similar taste to the user (Sarwar et al., 2001). On the other hand, content-based filtering uses item attributes to recommend other items similar to the user's preferences depending on the previous data points. The recommendation system can also combine both content and collaborative filtering by taking advantage of both techniques to recommend items (Van Meteren and Van Someren, 2000).

A few standard techniques for recommendations are mentioned. However, in reality, companies do extensive research to make recommendations as good as possible. Nevertheless, this book does not focus on complex recommendation systems.

8.1.2 Predictive Analytics

Predictive analytics are techniques to predict future values based on current and past events and data analysis. It exploits historical data to find patterns, applies these patterns to incoming data, and provides a predictive score for items and individuals depending on the context. For example, predictive analytics can estimate if a customer would churn at the end of the contact or score an individual for credit (Provost and Fawcett, 2013).

Predictive analytics make use of regression and machine learning techniques. Regression models are the workhorse of predictive analytics. The focus of regression is to establish a mathematical formula to predict an outcome based on different variables. Given an input, an output is estimated with the derived formula from past events. For example, it can estimate if the customer can pay for an additional service based on variables like location, salary, and services used. Hence, the sales representative can avoid reaching out to customers who would not accept the offer. Some of the common regression methods are linear regression, discrete choice models, and logistic regression.

In linear regression, a model forms between a dependent variable (the value we are trying to predict) and explanatory variables. The relationship is modeled through linear predictor functions where each explanatory variable gets a coefficient to compute the estimated value for the dependent variable. For example, the budget needed for a cloud solution based on the usage patterns can be computed. In this formula, dependent variables can be the number of engineers and the amount of data.

Discrete choice models can help to predict preferences between one or more options. If the user is browsing through the website's pages, it can predict if the user is looking for daily clothes or fashionable clothes based on the pages s/he browsed. When the user returns to the website, a recommendation is provided based on the user's preferences.

In logistic regression, the probability of a certain event happening is estimated, e.g. pass/fail. Each object will get a probability between 0 and 1. If a user's probability on vacation can be predicted, specific activities can be suggested for them to explore, or the probability of a transaction, fraudulent or not, can be expected. The services are aligned to reflect based on the probability value.

Consequently, the value for an observed event based on the model created by processing offline data with predictive analytics can be predicted. Predictive analytics can reduce cost and provide opportunities to improve user experience. Note that we kept our list short. However, many other methods, such as time series or decision trees, can be used for predictive analytics.

8.1.3 Pattern Discovery

Pattern discovery tries to find out items and structures that are strongly correlated in a given data set. It uncovers regularities in the data sets, relies on unsupervised learning methods to discover patterns where minimal human supervision is needed, and answers questions like what products should we offer together, how should we organize goods, and which areas should we invest in.

Many different techniques, like similarity matching, clustering, and co-occurrence grouping, can be employed. Similarity matching tries to find similar individuals or items based on the data. For example, a company can try to find customers that are close to their current customer base. Clustering can group the population into a different set of categories. An organization might offer different services based on the clustering information. Co-occurrence grouping helps to determine the association between items based on transactions. A supermarket might organize its isles based on the co-occurrence of products in transactions (Provost and Fawcett, 2013).

8.2 Data Science Life Cycle

The data science life cycle is a relatively new concept. It is an extension of the data life cycle, which has a long history in data mining and data management. A typical data life cycle has steps like data acquisition, data cleaning, data processing, data publishing, and saving results (Stodden, 2020). As an extension of the data life cycle, the data science life cycle has the same phases with additional steps. It consists of business objectives, data understanding, data ingestion, data preparation, data exploration, feature engineering, data modeling, model evaluation, model deployment, and operationalizing. The data science life cycle is depicted in Figure 8.1.

8.2.1 Business Objective

The vital part of creating a data solution is to establish the grounds for the work. It sounds obvious. However, many people rush into solutions without thinking

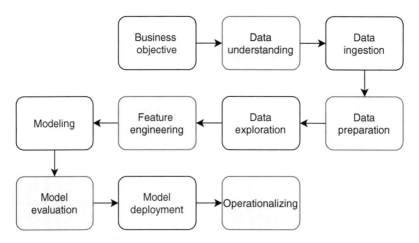

Figure 8.1 Data science life cycle.

about business value. Once a business problem is identified, it can be cast into potential data science problems. The following questions require answers:

- How many items can we sell?
- How much should be the price of the new item?
- How should we arrange our items?
- Which segment of the customers should we target?
- Are spikes normal?
- Which books should we recommend?
- What user groups do exist?
- Is it likely for customers to accept a new offer?

These are a couple of questions that need data science for answers. Nevertheless, it is still not enough to have questions. Justification on how the organization benefits by investing in the data science project is needed. The project's objective might be customer retention, additional revenue, and user engagement. It is crucial to set the objective to start the project.

8.2.2 Data Understanding

Understanding data is a prerequisite to building a data science solution. However, teams do not accommodate enough time for it. The meaning and the strengths, costs, and weaknesses of the data should be understood. The historical data might not be comprehensive for the problem being solved. Moreover, it may be slightly irrelevant to the solution. Even if the data is available, there might be costs associated with it.

Historical transactions or logged data may not have concerns similar to data science solution. The data might lack the information needed or might be in a different format. Revisions of the infrastructure might be needed to get better data. Moreover, the data might have different issues like reliability and consistency. While building the solution, several trade-offs are made about those issues.

There is also an associated cost to retrieving data. Buying the data from a third party company or partner with another organization to bring in data might be needed. Even if the data is in the platform, there is some preprocessing cost. While thinking about a data solution, the costs should also be considered as it could be challenging to justify the expenses later. It is important to uncover costs before investing.

8.2.3 Data Ingestion

Data ingestion might be a step to skip if the data needed is in the storage system. In reality, additional data or modifications to current data integrations are often needed, such as integration with a completely new data source - a new database, third-party data. If the Big Data platform is flexible enough, integration to a new data source may be easy. There will be more focus on frequency and backfilling once the integration's technical part has been figured out. A daily update to the data might be good enough for many cases, but it also depends on our deployments' frequency. Moreover, there is a need to backfill historical data. Backfilling can occur when a modification is needed in the existing pipeline, and extra attributes are computed down the chain.

8.2.4 Data Preparation

Once all data is in the storage system, it is ready for data preparation. Preparation involves formatting, removing unrecoverable rows, consistency modifications, and inferring missing values. Other than structural issues, information leaks should be observed.

The source format of the data might not fit in the project. It can come in JSON and should be converted into a column-oriented format. If some rows are corrupted or missing many values, it might be a good idea to remove them. Consistency problems like uppercase/lowercase or yes/no vs 1/0 can be addressed. In the last step, missing values can be inferred through imputation. Imputation is the process of replacing missing values in rows with substitutions (Efron, 1994). It can use different techniques like mean/median, most frequent, and feature similarity.

A leak is a condition where collected data has columns that give information about the target but cannot be available at the computation (Kaufman et al., 2012). For example, there is a total number of items in the shopping cart at the checkout time. However, the total number of items can only be known after the checkout. Therefore, it cannot be part of the model when the customer is still shopping. Leakage should be carefully taken as modeling happens in historical data.

8.2.5 Data Exploration

Data exploration is a phase to form a hypothesis about data by unveiling points of interest, characterizations, and primary patterns. Data exploration can use many tools like data visualization software to libraries like pandas. Data exploration tools help summarize data by simple aggregations to uncover the nature of the data and provide an easy way to detect the relationship between variables, outliers, and data distribution.

Data exploration tools can be used to script many aspects of the data. With automated exploration tools, the data can be visualized in many ways. Variables can be chosen, and statistics derived from them. Moreover, many visualization tools allow writing SQL statements where a more sophisticated view of data can be leveraged. Consequently, data insights that can inspire subsequent feature engineering and later model development are sought.

8.2.6   Feature Engineering

Feature engineering refers to all activities carried to determine and select informative features for machine-learning models (Amershi et al., 2019). A feature is a direct or formed attribute in a data set. Feature engineering generally involves adding new features based on existing ones or selecting useful ones from a data set and helps data science algorithms perform better. It is an indispensable element that makes the difference between a good model and a bad model.

Feature engineering requires a good grasp of domain knowledge. It often requires some creativity to come up with good features that can potentially help algorithms. Suppose the business of takeaway restaurants and their delayed orders need to be examined. There are columns like order id, the timestamp of order, the number of items they ordered, and the delivery timestamp. The dataset can be enriched with two additional columns. The first one is to add another attribute, such as order hour, that depends on order timestamp. The second one is the order duration that depends on the order timestamp and delivery timestamp. With these two additions, the hour of the day is correlated easily with durations, and the delivery time for a customer is predicted.

There is no one size fits all solution for feature engineering as it heavily depends on the domain. Common approaches still need work. The columns can be generalized such that it is easier to fit the model. Instead of age, a feature like age groups can be added, and sparse categories with *others* can be replaced to get better results. After adding new features, redundant columns created from new features and unused features like order id can be removed, as they do not provide anything for a potential model. Lastly, data science algorithms generally do not have a way of dealing with enumeration types, which can be replaced with numerical values.

It takes time to engineer good features and sometimes engineered features might not help the model as much as we expected. The features need to be revisited and revised. Feature engineering is not popular, but it is fundamental to building good models. It supplies features that algorithms can understand.

8.2.7   Modeling

Modeling is the step where data science algorithms come to the stage. In the first section, data science applications using algorithms have been discussed. One or more algorithms are chosen for modeling. The models chosen to train depends on several factors like size, type, and quality of the data set. When models are chosen, duration, cost, and output should be monitored. Modeling consists of several steps, as follows:

- Choose one or more algorithms such as random forest and decision trees.
- Split the data into two sets. The first split is for training purposes, and the second split is for model testing.
- Feed data into the chosen algorithms to build the model.

Once the model is built from the data set and algorithms, the next step is to evaluate the performance of the models. Problems may arise during the modeling stage, and there will be a need to go back to feature engineering or even further to get better results.

8.2.8 Model Evaluation

Evaluating models is a decisive step for the data science life cycle. Depending on the evaluation outcome, deploying a model, or re-engineering, previous steps can be done to revise the current model. During model training, the chosen models are trained with a selected validation method. The model with predefined metrics is tested in the model evaluation.

Several validation methods split the data set, such as hold-out, cross-validation, leave-one-out, and bootstrap. The available data are separated into two mutually exclusive sets in the hold-out method as a training set and a test set. The model on the training data set is trained and evaluated with the test set. In the k-fold cross-validation method, the dataset split into k mutually exclusive data subsets, k-folds. One of the subsets is used as a testing set and trains the model on the remaining sets. The process is repeated k times, each time choosing a different test set. One data point is left out for testing and trains on the remaining data in the leave-one-out method. This is continued until all data points get a chance to be tested. In the bootstrap method, m data points are sampled from the data set. The model trained and tested against it. The process is repeated m times (Kohavi et al., 1995). Depending on the amount of the data and compute resources available, a validation method is chosen. The hold-out method is cheaper and easy, but the rest require a bit more sophistication and compute resources.

There are various ways to check the performance of the model on the test data set. Some of the notable measures are confusion matrix, accuracy, precision, sensitivity, specificity, F1 score, PR curve, and receiver operating characteristics (ROC) curve (Powers, 2011). A few measurements will be briefly discussed. Recall the concept of a confusion matrix as follows:

- *True positives (TP)*: Predicted positive and are positive.
- *False positives (FP)*: Predicted positive and are negative.

- *True negatives (TN)*: Predicted negative and are negative.
- *False negatives (FN)*: Predicted negative and are positive.

Accuracy is commonly used and can be defined as follows.

$$\frac{TP + TN}{TP + FP + TN + FN} \tag{8.1}$$

Precision is another measurement method that tries to answer how much the model is right when it says it is right. The formula for precision is as follows:

$$\frac{TP}{TP + FP} \tag{8.2}$$

The measurement method depends on the domain. For example, false negatives for cancer should not be missed. Thus, a measurement method that aligns with such concern is needed.

8.2.9 Model Deployment

After evaluating the models, a model that meets expectations and has a reasonable performance is selected. The selected model is placed into a real-world system where investment can finally be returned in the deployment step. Depending on the solution, the model's interaction with the outside world may vary. A classifier may expose an API for the outside world to make classifications.

Typically, a model can be serialized into various formats, and the serialized artifact can be loaded back into production with the rest of the system. Once the model gets loaded, it can receive new data and return results. If the model is expected to receive real-time requests, a Rest API can be built around the model and service requests. Although the manual deployment of the model is possible, it would be highly valuable to have an auto-deployment mechanism to enable data scientists to train new models easily and deploy their solutions on demand.

When building a data science solution, a task queue can be useful. Machine learning models may take some time to evaluate a given request. Therefore, distributing work through a queue to multiple workers and letting the caller return immediately instead of waiting for the result can solve potential delays. The calling service can poll for the status and get the result once the model evaluates the result, and the worker saves the result to the result storage. With the task queue, the deployment might look like in Figure 8.2. Nevertheless, the suggested solution increases the complexity of the system by additional components. Thus, the requirements should be carefully considered before committing to them.

8.2.10 Operationalizing

Once there is a successful working solution, the last step is to operationalize. Operationalizing requires automation, testing, and monitoring practices. Most

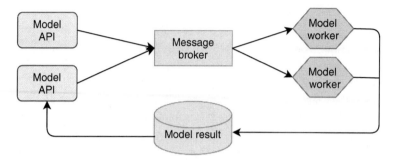

Figure 8.2 A sample data science model deployment.

data science projects start as a research and development project. At some point, they become part of the production system. Thus, they have to meet production system requirements as the rest of the infrastructure.

Operationalizing a data science project requires investment from the organization as they are different from the standard software development life cycle. A good way to handle operationalizing data science projects is to have people with operational experience in the data science teams. With this approach, the operational requirements can be baked into the solution early on. Deployment, monitoring, and other operations can become relatively easy. This will be discussed further down in the chapter.

8.3 Data Science Toolbox

Data science has a great set of tools to experiment, explore, and analyze data sets in many different directions. In this section, some of the well-known tools such as R, Python, SQL, TensorFlow, and SparkML will be discussed.

8.3.1 R

R is a language, environment, and platform for statistical computing, machine learning, and graphics. It is an integrated suite of software facilities that offer data handling, storage facility, collection of intermediate tools for data analysis, graphical abilities, an interactive environment, cutting-edge algorithm libraries, and a straightforward language with built-in commands for basic statistics and data handling. R's power comes from the library ecosystem (R, 2019).

R is a powerful data science tool due to the wide range of machine learning techniques through many libraries. It covers several techniques from data analysis, visualization, sampling, supervised learning, and model evaluation. R has cutting edge machine learning algorithms and procedures, since academia uses it day-to-day, and offers state-of-the-art techniques that may not be available on other platforms.

In this example, R is used to classify iris plant type. To do so, a library called *caret* that offers a set of functions for creating predictive models is used. The data is loaded and split into two half: test and training. The two models are trained with *k*-nearest-neighbor and random forest techniques. The test data values with *knn* are predicted, and the results are displayed.

```
# use library caret
library(caret)
# load iris dataset
data(iris)
# set the column names in the dataset
colnames(iris) <- c("sepal_length","sepal_width",
  "petal_length","petal_width","species")
# partition data into two sets, test(10%) and training(90%)
test_index <- createDataPartition(iris$species, p=0.9,
   list=FALSE)
# select 10% of the data for test
iris_test <- iris[-test_index,]
# use the remaining 90% of data to training
training <- iris[test_index,]
# select training data

training_features <- training[,1:4]
# select classes
training_classes <- training[,5]
# cv stands for cross validation, use 5-fold cross validation
control <- trainControl(method="cv", number=5)
metric <- "Accuracy"
# the seed has to be set to get consistent sequence
set.seed(11)
# knn stands for k-nearest-neighbor
knn_fit <- train(training_features, training_classes,
   method="knn", metric=metric, trControl=control)
set.seed(11)
# rf stands for random forest
rf_fit <- train(training_features, training_classes,
   method="rf", metric=metric, trControl=control)
# knn does a bit better so we choose knn over rf
predictions <- predict(knn_fit, iris_test)
# show results
confusionMatrix(predictions, iris_test$species)
```

8.3.2 Python

Python's popularity is ever-growing in data science. Python offers many libraries and great extension mechanisms. Libraries like *NumPy*, *SciPy*, and *matplotlib*, *pandas, scikit-learn* make it easy to address a wide range of data science problems. Pandas offers data analysis and manipulation tools. *Scikit-learn* offers machine learning techniques based on *NumPy, SciPy,* and *matplotlib*. Moreover, it is relatively easier to deploy Python implementations as there are web frameworks like Flask to offer machine learning capabilities through APIs.

In the following example, *pandas, requests,* and *tabulate* are used for exploratory analysis of COVID-19. A CSV file that contains cases, deaths, month, year, and places is first downloaded, a data frame is created out of the CSV file using *pandas,* and some of the columns are projected from the data frame that is grouped by projected values by places. The ratio between cases and deaths are identified, the data frame sorted in descending, and the top 20 values taken. The values are printed using *tabulate*.

```
import pandas
from tabulate import tabulate

url = "https://raw.githubusercontent.com/" + \
    "yusufaytas/datasets/master/covid19.csv"
covid = pandas.read_csv(url)
covid_summary = covid[['year', 'month', 'day', 'cases', \
    'deaths', 'countriesAndTerritories']]
covid_sum = covid_summary\
    [['countriesAndTerritories', 'cases', 'deaths']] \
    .groupby(['countriesAndTerritories']) \
    .sum()
covid_sum['ratio'] = covid_sum['deaths'] / covid_sum['cases']
covid_top_20 = covid_sum.sort_values(\
    ['ratio', 'cases', 'deaths'], ascending=False) \
    .head(20)
print(tabulate(covid_top_20, headers='keys', tablefmt='psql'))
```

In the next example, k-means clustering, which aims to divide data set into k clusters, is leveraged. k-means clustering finds clusters based on the Euclidean distance. A simple, practical use case for k-means clustering is usage segmentation, where there are metrics like session length and number of logins. With the help of k-means clustering, a business can find usage segments for its user base and offer a product range for each different segment. For the sake of simplicity, values are generated rather than referring to a data set and use *scikit-learn* to find clusters.

```
import matplotlib.pyplot as plot
from sklearn.datasets import make_blobs
from sklearn.cluster import KMeans

#number of samples
no_samples = 1009
#3 centers for session length in minutes vs number of logins
```

```
centers = [(47, 11), (37,3), (11,1)]
no_clusters = len(centers)

# Generate data
features, targets = make_blobs(n_samples = no_samples, \
    centers = centers, n_features = no_clusters, \
    center_box = (0, 1), cluster_std = 3)
# Remove negative values
features -= features.min(axis=0)

kmeans = KMeans(init='k-means++',
    n_clusters=no_clusters, n_init=5)
kmeans.fit(features)

# Predict the cluster for all the samples
predictions = kmeans.predict(features)

#  Generate scatter plot for training data
colors = list(map(lambda x: '#DA70D6' if x == 1 else '#FFD700'\
    if x==2 else '#7FFFD4', predictions))
plot.scatter(features[:,0], features[:,1], c=colors,\
    marker="o", picker=True)
plot.title('Usage Segments')
plot.xlabel('Session Length')
plot.ylabel('No of Logins')
plot.show()
```

The above code generates the plot in Figure 8.3.

The example is an introduction to machine-learning algorithms. Many models can be employed for classification, regression and, clustering. However, it will not be discussed in this book.

8.3.3 SQL

SQL is one of the most effective tools to cleanup, transform, and validate data. It also makes the exploratory analysis more comfortable. Some of the tools like Presto can also aggregate data in various ways. For instance, Presto has got approximate functions so that the queries sample data to speed up the analysis process. Moreover, Presto has statistical, metric, and entropy functions to support research in many dimensions.

8.3.4 TensorFlow

TensorFlow is an end-to-end open-source machine learning platform that operates at a large scale in heterogeneous environments. It abstracts the complex

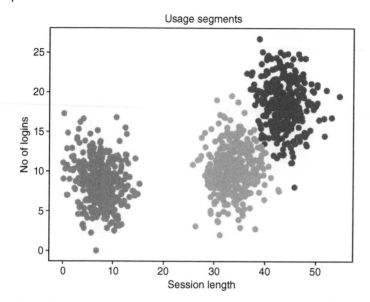

Figure 8.3 *k*-means usage segments.

machine learning computation algorithms and provides an easy API to create machine learning models for desktop, mobile, web, and cloud. It has a rich ecosystem of tools, libraries, and documentation and helps researchers develop state-of-the-art machine learning techniques and easily build and deploy machine learning applications. It uses data flow graphs to represent computation, state mutation and shared values and map nodes of data flow graphs to many machines (Abadi et al., 2016).

TensorFlow is available in multiple platforms and languages, including JavaScript, and it does not need to understand the underlying framework much to build applications. The dataset can be trained with minimal understanding. An example is a famous Boston housing dataset containing housing information such as crime rate and the average number of dwelling rooms. The problem requires predicting house value based on the properties of the house and neighborhood.

```
from tensorflow import keras
from tensorflow.keras import layers
from pandas import read_csv
from sklearn.model_selection import train_test_split
# load the dataset
url = "https://raw.githubusercontent.com/" + \
    "yusufaytas/datasets/master/boston_housing.csv"
boston_housing = read_csv(url, header=None)
```

```
# split into features and targets
features, targets = boston_housing.values[:, :-1],
boston_housing.values[:, -1]
# split into train and test datasets
training_features, test_features, \
    training_targets, test_targets \
    = train_test_split(features, targets, test_size=0.19)
# define model
model = keras.Sequential(
    [
        layers.Input(shape=(training_features.shape[1],)),
        layers.Dense(11, activation="relu", \
            kernel_initializer='random_normal'),
        layers.Dense(7, activation="relu", \
            kernel_initializer='uniform'),
        layers.Dense(3, activation="relu", \
            kernel_initializer='he_normal'),
        layers.Dense(1, name="last_layer"),
    ]
)
model.compile(optimizer='adam', loss='mse')
# fit the model
model.fit(training_features, training_targets, \
    epochs=499, batch_size=16, verbose=0)
# evaluate the model
error = model.evaluate(test_features, test_targets, verbose=0)
print('MSE: %.3f' % error)
# make a prediction
features = [0.03537,34.00,6.090,0,0.4330,6.5900,40.4, \
    5.49171,329.0,16.10,395.75,8.5, 10]
result = model.predict([features])
print('Predicted: %.3f' % result)
```

First, the data is downloaded and split into training and test data sets. Then a sequential training model is created with the Keras library. A sequential model groups a linear stack of layers. Each layer is a neural network where automatically infer prediction rules. The training set is fitted into the model, and then the results are tested with the test set. In the last part, a value is predicted. At this point, the model can be exported and used in different environments, even on the web, through JavaScript.

8.3.4.1 Execution Model

A TensorFlow graph consists of edges representing input/output to a vertex and vertices representing a local computation unit. TensorFlow computations at vertices are called operations, and values that flow through the graph are called tensors.

Tensor A tensor is a multi-dimensional array where each element is a primitive type such as *int32*. All tensors are immutable and hold identical data type with a known shape. A tensor can originate from an input data or the result of a computation.

Operation An operation takes m tensors as input and produces n tensors as an output. An operation has a named type such as *MatMul*.

Execution TensorFlow is deployed as a collection of tasks on a cluster where a task is a named process with a graph execution API. A task consists of one or more devices, and a device is an abstraction that corresponds to the CPU, GPU, or TPU (Tensor Processing Unit). Devices execute operations such as matrix multiplication through kernels. A kernel is an implementation of an operation. TensorFlow supports registering multiple kernels for the same operation to target different devices such as GPU and CPU easily. Moreover, a single kernel implementation can be compiled against a different device or data type to improve performance.

TensorFlow represents all possible computations via the data flow graph. The data flow graph expresses communications between computations explicitly. Therefore, it enables computations independently. Independent computations (subgraphs) make it possible to distribute the execution of large flows and parallelize the work to multiple devices. TensorFlow runtime places operations on devices subject to the constraints such as colocation and executes parts of the data flow graph in parallel.

TensorFlow offers tuning and failure recovery. TensorFlow runtime allows expert users to tune the placing of operations on different devices to optimize performance and deals with failures as training a model can take several hours to complete. TensorFlow addresses failures through a checkpointing mechanism where it saves the state to a distributed file system.

8.3.4.2 Architecture
TensorFlow architecture has several layers. On the top layer, there are training libraries and language clients such as Python client. The top layer talks to the rest of the TensorFlow through a C API. The core TensorFlow library is implemented in C++, and it can run in many operating systems such as Windows, Linux, and macOS. TensorFlow has several components like a distributed master, data flow executor, and kernel implementations. The distributed master converts client requests into task execution. The data flow executor accepts a request from the master and schedules a kernel execution that runs on a particular device such as CPU or GPU. The architecture overview is illustrated in Figure 8.4.

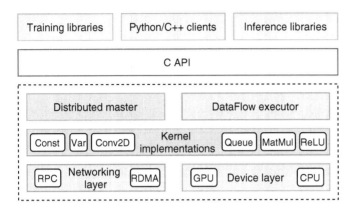

Figure 8.4 TensorFlow architecture.

8.3.5 Spark MLlib

Spark MLlib is the machine learning library that leverages the Spark execution engine to train machine learning models. MLlib covers a wide area of machine learning algorithms such as clustering, classification, recommendation, and regression. Since Spark offers other capabilities such as extraction and transformation, it becomes a powerful tool to combine with the rest of the Spark environment. An example is using Spark MLlib in recommending movies for users.

```
import org.apache.spark.ml.evaluation.RegressionEvaluator
import org.apache.spark.ml.recommendation.ALS
import org.apache.spark.SparkFiles
import org.apache.spark.sql.types.StructType
import org.apache.spark.sql.catalyst.ScalaReflection

case class Rating(userId: Int, movieId: Int, rating: Float,
    timestamp: Long)
val ratingSchema = ScalaReflection.schemaFor[Rating].dataType
    .asInstanceOf[StructType]

spark.sparkContext
    .addFile("https://raw.githubusercontent.com/yusufaytas"+
    "/datasets/master/movie_ratings.csv")

//read CSV file with ratingSchema
val ratings = spark.read.format("csv")
    .option("header", "true")
    .schema(ratingSchema)
```

```
    .load(SparkFiles.get("movie_ratings.csv"))
    .toDF()
val Array(training, test) = ratings
    .randomSplit(Array(0.8, 0.2))

// Build the recommendation model using ALS on the
//   training data
val als = new ALS()
    .setMaxIter(7)
    .setRegParam(0.01)
    .setImplicitPrefs(true)
    .setUserCol("userId")
    .setItemCol("movieId")
    .setRatingCol("rating")
    .setColdStartStrategy("drop")
//we need it to avoid NaN for evaluator


val model = als.fit(training)
val predictions = model.transform(test)

val evaluator = new RegressionEvaluator()
    .setMetricName("rmse")
    .setLabelCol("rating")
    .setPredictionCol("prediction")
val rmse = evaluator.evaluate(predictions)

println(s"Root-mean-square error = $rmse")

// Generate top 10 movie recommendations for each user
val movieRecommendations = model.recommendForAllUsers(10)
movieRecommendations.show()
```

Spark ML can be used to find movie recommendations for a user. A CSV file that contains user id, movie id, rating, and timestamp of the rating is read. Once the CSV is ready, it is mapped to a rating schema. An ALS (alternating least squares) algorithm is created for the recommendation, and the dataset is split into training and test. The model is trained using ALS, the error rate is checked, and the recommendations are presented to the users.

In addition to Spark DataFrame API, Spark comes with a high-level API, ML Pipelines, to build practical machine learning pipelines with Spark. ML pipelines work in a series of stages that are either a transformer or an estimator. A transformer is an abstraction where we implement a transformation function such as adding a new feature. The estimator is the model training abstraction that fits the training set to the model. A transformer implements a method transform, and an estimator implements a method fit. A pipeline accepts an array of transformers and estimators to sequentially apply them in stages. In the following example, we use a series of transformers and a logistic regression form the pipeline.

```
from pyspark.ml.classification import LogisticRegression
from pyspark.ml.feature import StringIndexer, VectorAssembler
from pyspark.ml import Pipeline

training = spark.createDataFrame([
    (0, 'f', 23, 1.0),
    (1, 'm', 32, 0.0),
    (2, 'f', 19, 1.0),
    (3, 'm', 43, 0.0)
], ['id', 'gender', 'age', 'healthy'])

#maps gender column to indexes
indexer = StringIndexer().setInputCol('gender')\
    .setOutputCol('gender_index')
#assembles features into a vector column
vectorAssembler = VectorAssembler()\
    .setInputCols(['gender_index', 'age']) \
    .setOutputCol('features')
#assembles features into a vector column
lr = LogisticRegression().setFeaturesCol('features')\
    .setLabelCol('healthy')
pipeline = Pipeline().setStages([indexer, vectorAssembler, lr])

model = pipeline.fit(training)
```

8.4 Productionalizing Data Science

Data science is a continuous research and development area for organizations. If efforts do not go beyond prototypes and models, the investment will dull. The methods to productionalize machine learning endeavors are needed to generate business value. Pushing machine learning models to production has been a relatively new effort for smaller or mid-size organizations. Recently, there has been some great work to fill this gap. In this section, some examples of machine learning deployment frameworks and software such as Apache PredictionIO, Seldon, MLflow, and Kubeflow are discussed.

8.4.1 Apache PredictionIO

Apache PredictionIO is an open-source machine learning server that allows engineers and scientists to create solutions to common machine learning tasks such as classification, regression, and recommendation. PredictionIO makes it easy to deploy a machine learning solution to production as a service. PredictionIO can respond to different machine learning requests once it gets deployed. PredictionIO can update its model in real time or batch (PredictionIO, 2019).

8.4.1.1 Architecture Overview

PredictionIO has three components: event server, prediction server, and training. The event server accepts a new event and stores it. Alternatively, the event server can store imported events. The event server uses Postgres for small data and HBase for a larger data set. PredictionIO uses Spark to train machine learning models. Many algorithms are available from Spark MLlib in PredictionIO. The prediction server accepts prediction requests from clients and responds using trained models. Models get shared through HDFS between different components. Moreover, PredictionIO uses HDFS for import/export from the event server. On top of HDFS, PredictionIO uses ElasticSearch to store mappings, evaluation results, and additional metadata (Figure 8.5).

8.4.1.2 Machine Learning Templates

PredictionIO has an extensible mechanism where it can have many applications running on the same server. An application has a registered engine template. An engine template is an implementation of a machine learning algorithm such as collaborative filtering. PredictionIO has a template gallery where the administrator can choose a template from. The administrator needs to deploy the engine through a configuration. Later, the administrator can then bulk import data or start event publishing through the API endpoint. Once the data is ready, the administrator can run the command where it would train the models. The administrator can schedule training periodically. Therefore, training does not have to be a manual process at all.

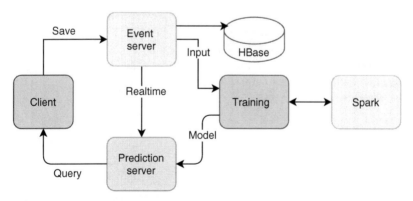

Figure 8.5 Apache PredictionIO architecture.

8.4.2 Seldon

Seldon is a software framework to make machine learning models deployable through microservices. It provides ways to containerize machine learning models into production-ready deployable artifacts. Using language wrappers, a model can conform to a structure that enables calling it externally through Rest API. Seldon integrates with popular monitoring solutions like Prometheus to monitor and statistics for the service out of the box. Seldon workflow has three steps, as follows:

- Wrap model with language wrappers or inference servers.
- Define the inference graph and deploy the service.
- Call service with API or client libraries.

Seldon allows data scientists to wrap their saved models easily as follows:

```
from sklearn.externals import joblib
class IrisModel(object):
    def __init__(self):
        self.model = joblib.load('iris_model.sav')
    def predict(self,X):
        return self.model.predict_proba(X)
```

In this example, the previously trained model is wrapped into a model class. The wrapper can be containerize using the Seldon core utils, *s2i*. Once the image is built, the model can be deployed to the kubernetes cluster. Seldon supports OpenAPI specifications. Therefore, the model will become a Restful endpoint to predict models as follows:

```
curl -X POST http://<ingress_url>/seldon/<namespace>
    /<model-name>/api/v1.0/predictions \
    -H 'Content-Type: application/json' \
    -d ' "data":  "ndarray":  [[1,2,3,4]]  '
```

8.4.3 MLflow

MLflow provides a solution for the data science life cycle. MLflow platform enables data scientists to package models into reproducible runs, register models, and monitor various phases of the development lifecycle. MLflow offers lightweight APIs that can be used with most machine learning libraries. MLflow has four components: MLflow tracking, MLflow projects, MLflow models, and MLflow model registry.

8.4.3.1 MLflow Tracking

The MLflow tracking component provides a user interface and API to keep track of machine learning experiments. It stores information about parameters, code versions, metrics, and output files for machine learning models. MLflow tracking has a concept of a run. A run is related to the execution of a machine learning model. In a typical run, MLflow tracking logs parameters, code version, timestamp, and artifacts. MLflow can store run information to a database through SQLAlchemy and a local file system. The user interface allows data scientists to compare and contrast different runs. Experimenting with the same model over different parameters and comparing them saves time for data scientists.

8.4.3.2 MLflow Projects

The MLflow Projects standardize the packaging of data science code. It allows for creating reusable and reproducible artifacts. At the very base level, MLflow projects define a project structure that organizes and describes code such that automation tools and data scientists can run experiments without knowing the details of the model. MLflow project defines project directories and the environment for machine learning models. It has an MLproject file that defines entry points for the machine learning model and the environment, such as conda or docker. An example project file is as follows:

```
name: dbdp prediction
conda_env: dbdp_conda.yaml
# we can define  a docker_env instead of a conda_env as
# docker_env:
#    image:  dbdp-docker
entry_points:
  main:
    parameters:
      data_file: /dbdp/input
      regularization: {type: int, default: 0.11}
    command: "python dbdp_prediction.py -r regularization
        data_file"
  validate:
    parameters:
      data_file: /dbdp/validation
    command: "python dbdp_validate.py {data_file}"
```

8.4.3.3 MLflow Models

The MLflow models standardize the packaging of the machine learning model. The MLflow allows downstream tools to call machine learning models easily. It defines multiple flavors that a model can be viewed. Flavors define how a model runs. For example, a python function flavor tells that the model can run as a python function. MLflow models have MLmodel files and arbitrary files. In the MLflow file, the flavors are described along with additional information.

```
time_created: 2020-09-14T20:28:53.31
#two flavors are defined
flavors:
  sklearn:
    sklearn_version: 0.23.2
    pickled_model: model.pkl
  python_function:
    loader_module: mlflow.sklearn
```

8.4.3.4 MLflow Model Registry

The MLflow model registry centralizes model management, where data scientists can view models in a central location. The registry provides lineage information and transitions between staging and production, and model information about versioning, descriptions, and tagging.

8.4.4 Kubeflow

Kubeflow is another machine learning automation tool that aims to deploy machine learning models scalable and straightforward on kubernetes. One of the key integrations of Kubeflow is the Jupyter notebook. Jupyter notebook is a web application that allows data scientists to create and share machine learning findings that contain live code, visualizations, and explanations. Kubeflow makes it easy to run multiple experiments in parallel for Jupyter notebooks. Kubeflow user interface offers easily launch new notebooks by choosing a Docker image for the Jupyter. Data scientists can set resources for the notebook, such as GPU or CPU.

Moreover, Kubeflow eases distributed training to get results faster for PyTorch and Tensorflow. Kubeflow offers custom resources for PyTorch and Tensorflow to easily run experiments with distributed models. Kubeflow offers additional components such as pipelines, metadata, and katib, to better cover the data science life cycle.

8.4.4.1 Kubeflow Pipelines

Kubeflow pipelines offer an intuitive user interface for managing and tracking machine learning experiments. Apart from the user interface, Kubeflow pipelines have an engine to schedule multi-step pipelines and an SDK to define and manipulate pipelines. Kubeflow pipelines allow the orchestration of containerized machine learning tasks. Each machine learning task, a step in the pipeline, takes input arguments and produces output files. The pipeline specification comes through a Python DSL. We can define pipeline input data and get the output visually, such as ROC curve or confusion matrix.

8.4.4.2 Kubeflow Metadata

Kubeflow metadata is a central repository to publish information about runs, models, datasets, and other artifacts. Kubeflow metadata comes with an SDK to record metadata.

8.4.4.3 Kubeflow Katib

Kubeflow katib is a tool for automated tuning of the machine learning model via hyperparameters. Hyperparameters are parameters used to control the learning process, such as the number of layers in the neural network, the number of nodes in each layer, and the number of trees in the random forest. Hyperparameter tuning involves optimizing a specific metric by adjusting hyperparameters for the model. Katib executes several trials simultaneously within different settings for each trial. Once the trials finish, Katib compiles the result and gives the optimized values for the hyperparameters.

9

Data Discovery

After reading this chapter, you should be able to:

- Reason about the importance of data discovery
- Learn data governance practices
- Understand tools for data discovery

Data discovery is a complex concept that could mean generating insights from different data sources to any sort of business intelligence activity. Nevertheless, my use of data discovery has a different implication. Data discovery makes metadata easily accessible, documented, and presented in different ways. It is a key piece since it enables the rest of the data platform activities. When the data is not discoverable, it generates friction in consumption. The better documented, available, searchable the metadata, the quicker the extraction of data.

This chapter discusses the importance of data discovery for a modern Big Data platform, find the link between data discovery and data governance, and explore tooling used in Big Data discovery.

9.1   Need for Data Discovery

The phenomena of data warehouses and later data lakes have been making data available in many forms. The quantity of data piped into the data lakes creates confusion, inconsistency, and errata. Since data are stored without requiring a structure, it becomes inherently harder to keep metadata consistent and meaningful. In big organizations, the problem becomes more challenging because numerous teams work in different departments using different tooling but shared infrastructure.

Designing Big Data Platforms: How to Use, Deploy, and Maintain Big Data Systems,
First Edition. Yusuf Aytas.

For a typical data lake, data sources are staggeringly diverse. Some of the data come from production databases. Some of the data come from CRM tools. Perhaps, some of the data get crawled. Due to simultaneous pumping, tables, schemas, and columns can become confusing. A conversion column might mean different things to sales and marketing. When there is no control over how things are named, or there is no way to access a data dictionary, the meaning of columns and related terms can become confusing to a data engineer evaluating the same set of tables. When there is no structure at all, metadata information can become a nightmare, and classical pipeline tooling will fail to impress.

When there are many teams with data processing needs, the teams can have their preferences for tooling. Some of them might just prefer a couple of cron jobs, whereas others would prefers tools like Airflow. In some scenarios, the teams can even build pipeline tools. The concern here is the representation of classification, lineage, and ownership of data. Since there is no central mechanism for tooling, there is no way to check data processing phases. A data analyst would find if confusing and challenging to trace the data if they are partially assessed, as well as to identify which team is involved or what category the data belongs. Even if the analyst eventually learns the business logic and phases, it would not be sustainable long term.

Regardless of the organization, people come and leave. When there is a huge learning curve, people will resist, adapt slower, and enjoy their job much less. Everyone wants to get the job fast, but they will fail to do so if there is so much to hinder them. If the data is not discoverable, people have to discover themselves, note it down personally, and maybe luckily to a shared document in the organization. Although the information might be partially available, it will be useful to some extent. When the person leaves and another person takes the job, it will start all over and has to go through each table, understand the process, and so forth. On the other hand, with a centralized metadata discovery, none of this might be necessary.

There are many other causes of inconsistencies and erasure in the metadata, such as organization changes, company mergers partner integration, and so on. I would like to discuss some aspects that make it necessary to have data discovery for a modern Big Data platform. Next, I would like to discuss various aspects of data discovery.

9.1.1 Single Source of Metadata

Disparate data sources in Big Data systems make accessing metadata troublesome. There is no easy way to track, memorize, and reason about metadata in subsystems. Besides, the underlying Big Data storage systems or even processing systems might differ a lot. The same column might have different types in the two systems. A metadata repository can capture these differences, definitions, comments, and more, and it is a central location for cataloging organization-wide data. We can leverage the metadata repository to retain current and historical trends of the same or similar metadata. With many more data sources integrating into Big Data platforms, the metadata repository has become a necessity.

A single source of metadata makes metadata available for further processing. It is impractical to expect people to query metadata repository directly. Nevertheless, a metadata repository is necessary to allow further processing like indexing, grouping, etc. There are two integrations we can do to centralize metadata. One integration point is to crawl the existing data warehouses, data marts, data lakes, etc., for metadata. What is more, we can crawl downstream database systems. The second integration point is to have an API in which other systems can register their data. An API approach is particularly useful when we have different engines such that they can register, update their schema automatically.

A metadata repository should contain all the details about data. A typical metadata repository would contain all schemas and tables for a given data source. For each table or data structure, it should contain column name, type, comment, and so forth. Having additional information like partition columns for a given structure can make consumption more fruitful.

9.1.2 Searching

One of the key aspects of everyday development is searching. It is no different for a Big Data engineer or analyst. The need for powerful searching becomes more noticeable when someone joins the team. With many data sources and ETLs, it is not straightforward to find the necessary information. Once we can establish a central metadata store, we can then start indexing our metadata. The platform tooling should enable data enthusiasts to search through column types, comments, and more.

The vast majority of the indexable source would come from relational databases or parsed data saved to our data lakes. The rest can come from API integrations and further structure data. The search functionality should be able to scan through all of these data sources. When combining these disparate sources, a ranking methodology among search results would ease finding information faster. For example, we can rank the query results based on the usage patterns, e.g. frequently queried tables show up earlier than less queried tables. What's more, we can improve the search results based on the number of follows or likes if such functionality exists.

9.1.3 Data Lineage

Data lineage gives us information about the journey of the data. It tracks data from an origin and shows what and where happened to data throughout various transformation stops. Having data lineage readily available helps in understanding the whole process. The data lineage is partially available in workflow orchestration. In data flows, we have multiple steps of aggregations. From one step to another, we have joins, filtering, and many other transformations. Nevertheless, it still does not cover the whole picture.

Data lineage consists of multiple directed acyclic graphs (DAGs) where each node in the dag corresponds to a table or a data structure. Each node in the data might have a different type of structure. Some of them might be static tables, and others might be staging tables or structures that help us build complex aggregations. Although the lineage can start from an actual table, that is, almost not the case in practice. The source can be event stream, logs, or some parsed external input. Most of the data we process comes continuously. It is also important to include the same source, what generates it.

Providing data lineage in Big Data platforms can help discover patterns, concepts, and a wider spectrum of organizational metadata. In most organizations, a similar type of aggregations happens in multiple places due to coordination or organization problems. Data lineage can immediately reveal the full picture of all necessary data computation. It can help to apply the single source of truth pattern organization-wide.

The implementation of data lineage can get complicated. The lineage can partially be constructed from data pipelines. That is probably a good starting point but not enough to cover all data flow. One way is to enable API integration with data discovery tooling where updates can be reflected automatically.

9.1.4 Data Ownership

Responsibility and accountability are the driving factors for data ownership. If we know which team owns a set of tables or data structures, we can get more information from the owning team. More importantly, if there's an issue with the data quality, we can reach out to the owner for a fix. In a shared infrastructure, many teams would engage in cross-team data transformation activities. Having a data owner makes each transformation a controlled and monitored process.

Nevertheless, it is hard to establish ownership organization-wide. In many cases, tables, pipelines, streams, and other data structures might belong to the person who worked on it. As much as it is an organizational problem, it is also a discoverability issue.

To address the ownership issue in Big Data platforms, we should put two routines in place. The first thing is to establish a process to determine which data does not have an owner or just an individual. This can happen automatically with the tooling by reaching out to individuals to claim their data and set appropriate ownership information to the data. The second thing is to require any data to be created with an owner. If we can have both of these in place, one can discover many connections by just looking at the metadata of the data. Having ownership routines integrated is beneficial for organizational knowledge, maintenance, and interactive reasoning.

9.1.5 Data Metrics

Metrics are useful for any environment. We want to establish the same routine for the data. We would like to know quick essential metrics about the data. Nevertheless, it requires good integration with data discovery tooling. Some of the important metrics for a given table or data structure can be as follows:

- *Last update*: The time where the most recent update happened to the data.
- *Partition information*: If the data is partitioned, it is important to see which partition columns are used to join it with another table.
- *Disk space usage*: Disk usage can give an engineer a quick estimate about any table that can be built.
- *Retention policy*: How far the data kept is essential for historical analysis.
- *Storage type/format*: If the table is stored columnar or row fashion. If compression is applied, what sort of compression method is used?
- *Generating platform*: There might be many tools working simultaneously while generating the data. A link to the generating pipeline or generating tool might make finding more accessible.
- *Versioning*: If the metadata is versioned, the version information might be an important factor for determining downstream data.
- *The number of queries*: Gives information on how much this table is used. If the table is not used, it can be deprecated.

There can be many other useful data metrics that can be added to the list. These metrics give an engineer or analyst a chance to assess the overall picture of the data quickly. The harder part is collecting these metrics into a central location with a metadata repository. Again, we can use API integrations with metadata discovery tools such that tools can publish information about various metadata. Another method is to passively populate these metrics using underlying data storage systems.

9.1.6 Data Grouping

When we copy metadata information from multiple data sources, we might end up in a situation where we would have multiple copies of the same data. This is very common as we might be just pouring data into the data lake from various resources or transferring summary data to a data mart. The data might look quite similar on a base level, although the underlying data types or storage system might change drastically. Multiple metadata definitions in a metadata repository can be quite confusing.

One way to tackle the proliferation of metadata is to detect and group these tables or data structures. For example, we can have a table, called members, in a Postgres database. The same table can exist in our data lake in a reserved schema for copies from the member database. If we can have a mechanism that checks similarities between tables or data structures based on column and table names, we can group them into one. Instead of serving individual tables, we can serve the grouped version of the table, which might link to the tables underneath.

9.1.7 Data Clustering

Clustering data into data sets helps to find out metadata related to each other. A quick example is to have all revenue related tables clustered into a data set. Looking at the data set, one can dive into any of the revenue related tables. Nevertheless, it probably requires manual work as it is hard to understand which tables should be in the same data set by just looking at the metadata. Perhaps, a semiautomated clustering job can be built where the job will find potential data sets and suggest to the owners. Once the owner is happy with the set of the tables, then metadata can be connected through data sets. Removals and additions to the clustered metadata might enable further improvement.

9.1.8 Data Classification

Classifying data into different categories has been employed in organizations, even without data discovery. Such classification needs to be done because of security and compliance. Classification is still a must in Big Data systems. We must be careful with personally identifiable information (PII) and other variants. Hence, classifying those data sources, tables, and data structures is important for visibility, compliance, and security practices. We can go one step further and introduce tagging and other mechanisms to classify metadata into further categories.

Once we exercise the classification of metadata, we can then quickly discover metadata in multiple categories. For example, we can look for finance, 2020, and sales tables. If we had proper tagging on each of the tables, we would find out information within seconds. Moreover, we can leverage tagging for notification purposes. We might subscribe to certain tags and get notified for updates if there is a change in the tagged tables.

9.1.9   Data Glossary

A data glossary is a repository to capture terms related to the business or organization. In a large organization, it is hard to get everyone to speak the same language. To speak the same language, we need some rules. This is where a data glossary, also called a business glossary, comes into play. We can define terms under different categories. For instance, we can define accounts payable under the accounting category. When we refer to accounts payable, then it would imply the same meaning for everyone.

When we discover data, we would like everyone to come to the same conclusion for a given data. Therefore, an implementation to define a data glossary and associate it to columns or tables can make definitions much clearer and avoid potential confusion between two disparate data objects. Moreover, columns or tables with associated terms would become more trustworthy. It would encourage other data owners to define their terms and use them in their tables and columns. Having strong descriptions can provide a solid data dictionary over data set for existing and potential customers.

The data dictionary is a matrix of definitions for each data object in a data set. It is a way to communicate with stakeholders technically. Each attribute in the data object has a description and might potentially refer to a glossary term. A well-defined data matrix can help new customer acquisitions and keep the data set consistent when discovering a certain data set.

9.1.10   Data Update Notification

In big organizations, many teams depend on others. Although people try to communicate changes and updates, sometimes it is anyhow missed. If we have a multistage pipeline where one team depends on the other one, an upstream change might break downstream pipelines. Perhaps, someone from the downstream team will get paged, and a couple of actions might have to be taken under stress. If we can have a notification mechanism for changes, we do not have to worry about communicating updates. Any downstream pipeline owners can follow to set of tables they are interested in. When there is a change, then they can react to it before it becomes a stressful incident.

Operational updates are one part of the picture. The other part is information updates. For example, the table structure might be the same, but someone might update the documentation on it. If we are interested in that particular table, then we can receive updates on definitions. We might also contribute back and help to fix misinformation. There can be other useful notification cases, but the aim is to enable communication channels about metadata updates. This gets more important when the organization grows bigger.

9.1.11   Data Presentation

We have described many different aspects of data discovery but kept the presentation as the last one. The presentation of metadata is the base for discovery. All of the other aspects of discovery depend on a good presentation layer. The presentation layer should display all details we mentioned, such as lineage, grouping, metrics, clustering, classification, and more. A graph representation of tables might help in displaying lineage as well as other cases. Moreover, we should also have individual data representations where we can add more information to the metadata. A good presentation layer helps in completing the feedback loop for data discovery.

9.2   Data Governance

There is a close relationship between data discovery and data governance. Data discovery can be regarded as a specialized branch of data governance focused on making information available. Data governance practices relate to the entire business, but I only focus on Big Data platforms for this book. Hence, I would like to go over data governance in the context of Big Data. Data governance addresses the journey and presentation of enterprise data with policies and practices. Data governance aims to ensure data assets conform to standards defined by policies. In the context of Big Data, data governance takes another shape as Big Data systems are relatively flexible in accepting data in any format from any data source. I would like to go over the data governance briefly first and then transition to Big Data governance.

9.2.1   Data Governance Overview

Data has been an important asset for organizations. Organizations have put effort to utilize data in the best way possible. In efforts for utilization, organizations have adopted data governance. Data governance aims to ensure data meets business goals, complies with industry standards, and stimulates new opportunities. In reaching its goals, data governance requires organizational commitment and executive sponsorship (Panian, 2010). If both requirements are met, the rest is to establish a framework for data governance concerning data quality, metadata, data access, and data life cycle (Khatri and Brown, 2010).

9.2.1.1   Data Quality
Data quality describes how accurate, credible, useful data is throughout the organization. Data quality depends on the data itself. We might compromise data quality for some data sets. However, we might need almost absolute certainty for some

others. When it comes to measuring data quality, there are a couple of factors to consider, such as accuracy, completeness, trustworthiness, and timeliness. Data accuracy measures the correctness of data. Data accuracy helps in understanding how much of the data is correct for a given data set. Data completeness indicates how much of the required data is in the data set. Data trustworthiness denotes how credible the data source and the data itself. Lastly, data timeliness shows whether data is up to date for given conditions. For example, if all data we have is a day old, we can think the data is timely.

Data quality efforts require monitoring of the data journey from ingestion to transformations. For example, we would like to keep the amount paid accurately for an order table. This requires having the correct form of the data in the messaging layer, data warehouses, and databases. Moreover, data quality techniques might involve extra processing, like data cleansing and validation. We might accept a property address, but after validation, we might reject the proposal. We can even better guess the correct address after postprocessing and propose it to the customer or user via some sort of notification.

9.2.1.2 Metadata

We have already covered the importance of metadata and making it an integral part of a Big Data platform. We extend it and tell that metadata should be addressed by the organization. The data initially gets generated by the applications, websites, and other integrations. If we can enhance metadata information about every data object, then we have a better chance of developing actionable and applicable results. An important factor is to enforce standards for every database table, messaging object, etc. For example, it might be mandatory to briefly describe of each object along with an explanation for all columns and attributes.

Once we can establish an organization contract for data creation, we can successfully transfer the metadata information to consuming parties or a metadata store to have a holistic view of data. Furthermore, it is important to address enriching existing metadata. If a data object is used actively, it means there are parties interested in that. We can then ask consumers or teams to improve the documentation. Nevertheless, writing documentation is a mundane task, hard to get everyone on board.

9.2.1.3 Data Access

Organizations run many disjoint applications to cover their data needs. And data access is an orthogonal requirement for all of the various applications. Data access determine who has access to a particular object and who owns which data in the organization. Nevertheless, achieving data access to multiple data sources is a challenging task. Each data system has its access model, and data models within each system might need different access rights due to environment, nature, and so forth. Despite challenges, establishing data access rights is compulsory for any business for security and compliance. It is hard to balance what is best for agility vs. security.

Data should be available for consumers without much friction. At the same time, the organization should not leak information to both external and internal parties. Having a well-defined data access policy and automation can ease the burden and friction. With a good structure of who can access which data sources, we can automate granting access rights for applications and personnel. Nonetheless, it is hard to implement automation that can integrate with various data management systems.

9.2.1.4 Data Life Cycle

Many organizations do not have comprehensive knowledge about their data assets. There are many questions about the life cycle of data, such as where it lives, which data sources contain it, and whether it is up to date. Without close monitoring of the data, we can't answer these questions. Answering these questions can help in optimizing storage as well as efficiency wins for transferring data. Nevertheless, organizations have to control the life cycle. Again, this is a problem of convenience vs. friction. We want people to easily transfer and aggregate data within the organization, but we want to see how data is evolving.

9.2.2 Big Data Governance

Many additional factors make it harder to apply data governance to Big Data. Big Data inherently has higher volume, higher variety, and higher velocity. Independent processing of large volumes of data requires increased supervision over data quality, metadata, data access, and data life cycle. Besides, the combination of Big Data and business tasks may lead to a more frequent and higher level of risk for a data breach (Yang et al., 2019). Big Data governance depends on the data architecture and data source integration, and factors for data governance.

9.2.2.1 Data Architecture

Large organizations tend to have many teams working on Big Data. There might be numerous data lakes, data warehouses, and data marts. The disconnect within the organization can impede visualizing the overall architecture for Big Data. When the interaction between multiple systems is not clear enough, we might have data inconsistency, inaccuracy, and access problems.

The evolving nature of software development makes it harder to fully understand of the overall data architecture. Nevertheless, the layout of the pieces is an important factor in governing data in multiple systems. We should consistently update the components of a Big Data platform as we evolve from one phase to another. We should look at data from access, quality, and life cycle dimensions within each component to ensure systems are compatible. The differences between systems should be considered for each data movement. For example, we should not leak PII data to systems that have company-wide access.

9.2.2.2 Data Source Integration

Data source integrations are a critical part of Big Data platforms. With each integration, we face challenges in terms of data quality, security, and consistency. Although it might be straightforward to transfer from one data source to another, it is not always practiced correctly. Thus, we should create control and best practices for data transfer. When it is possible, we should automate some of the best practices so that we can remove the human error factor.

Most of the errors come from wrong setups for the transfer and transformations. If we can put guidelines for both, we can then avoid such problems. The guidelines should cover things like type conversion, mapping, and access control. We often take shortcuts for short term wins; however, such shortcuts devour the whole data infrastructure. Thus, we should not override integration guidelines or best practices for short-term gains. We might immediately accumulate the technical debt for the wrong reasons.

9.3 Data Discovery Tools

Processing Big Data requires contextual information. With the growing amount of metadata, some organizations came out with solutions to grapple with data discovery. Some of these tools have been open-sourced, and some of them are not. There are also proprietary solutions to address data discovery needs. Nevertheless, discovery tools do not address all the needs we discussed previously, but they still provide neat solutions. In this section, we will go through some of these solutions for data discovery.

9.3.1 Metacat

Metacat is a service that serves as a Rest/Thrift interface to access metadata with underlying data sources. The data sources provide the actual metadata information. Metacat integrates with them and provides an abstraction layer on top of them. While obtaining metadata information, Metacat publishes the metadata information to ElasticSearch for searching purposes. Metacat provides a centralized metadata repository, interoperability over multiple data sources, searching functionalities, update notifications. Metacat has an API layer, a service layer, and a connector layer. API layer provides unified information about metadata.

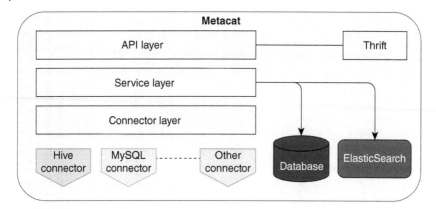

Figure 9.1 Metacat architecture.

The service layer saves metadata to a central repository and publishes it to ElasticSearch (Majumdar and Li, 2018). The connector layer connects to various data sources, as shown in Figure 9.1.

9.3.1.1 Data Abstraction and Interoperability

Metacat integrates with multiple query engines such as Pig, Spark, Presto, Hive, and so forth. Metacat introduces an abstraction for these query engines where data sets can be shared between engines reciprocally. For instance, a Pig Latin script can read Hive data types with the help of Metacat. What is more, Metacat can help the mapping of types from one system to another. When transferring tables, the mapping can be used to create tables in the destination. Metacat implements a thrift protocol to enable running queries over presto and spark.

9.3.1.2 Metadata Enrichment

In addition to the metadata itself, Metacat stores business or user-defined information about datasets in its storage. The additional information is free form. One can leverage it for statistics, metrics, partitions, and retention. We can put table completion times and ETL related statistics. We can add metrics like space and cost for the table. The cost for the table can be computed by the space it takes vs. the overall cost of the storage. For partitioned storage, we can put partition information along with retention. The table owners can also set column default values as well as rules for writes on the table.

9.3.1.3 Searching and Indexing

Since Metacat publishes all metadata information to ElasticSearch, indexed data can be consumed in many ways. With the help of the ElasticSearch, metadata can be searched easily. We can set filters or search directly with free text form. Since the data gets published to ElasticSearch, we can make additional uses like auto-completion or suggestions while writing queries.

9.3.1.4 Update Notifications

Metacat gets all the updates from each one of the systems. Having all metadata, it can send update notifications. Nevertheless, it only supports one cloud provider integration for the updates as of now. Additional development needs to be done to support different notification mechanisms.

9.3.2 Amundsen

Amundsen is an open-source data discovery and metadata engine. Lyft engineers developed Amundsen to increase the productivity of their data scientists, analysts, and engineers. When the amount of data increased over time, it becomes increasingly hard to find any information. Besides, compliance and regulatory challenges have to be addressed when processing Big Data. Amundsen provides metadata searching capabilities, contextual information about data, and ranking tables based on usage metrics (Grover, 2019).

9.3.2.1 Discovery Capabilities

Amundsen offers rich discovery capabilities. It quickly shows popular tables on the landing page along with a search box, where we can search for organizational data. It ranks search results based on the popularity, e.g. most queried tables come first. The search results provide some metadata such as a name, description, last updated, and so forth. Once we found the table or object we needed, we can then check out the details.

In the details, Amundsen gives rich information about the table. It shows each column description and table related details such as owner, lineage, frequent users, tags, links to source code, pipeline tool, etc. It can also sample data and provide quick exploration for the table. Users can update the description and tag information on the table. Amundsen makes basic information available for everyone in the organization. However, it restricts access to certain functionalities if the table has restricted access. Making all metadata available helps people in the organization to find out tables. If they need access to these tables, then they can request access to it.

Amundsen gives a couple of column metrics for an individual column, such as count, max, min, distinct count, null count. These stats are only available to users who have access to the table. Some of the tables can contain sensitive information; therefore, it has to have proper access control over such metrics.

9.3.2.2 Integration Points

Amundsen integrates with many popular tooling across the organization to help to discover data in many dimensions. Integration points provide a central location to access information about metadata, usage, processing, source, and ownership. Amundsen combines the following integration points to give a complete view of the data:

- Data storage systems such as Hive, Presto, and relational databases, object stores, and so forth.
- Dashboard and reporting tools like Apache Superset for exploration as well as sharing saved queries.
- Schema management systems to consume schema from schema registries.
- Stream information such as topics and streams from message brokers like Kafka.
- Data pipeline tools like Airflow to get ETL job information.
- Access management tools like LDAP to get organizational structure, name, title, access, etc.

9.3.2.3 Architecture Overview

Amundsen consists of a couple of services, a common library between these services, and a data ingestion library for the metadata graph. The frontend service powers up the user interface. The search service wraps ElasticSearch functionalities into service. The metadata service provides the persistency through Neo4j or Apache Atlas. Data ingestion library helps to connect to various data sources to gather metadata information as well as building a search index (Figure 9.2).

Amundsen Frontend Service The frontend service is the entry point of Amundsen. The frontend services provide user interface functionality to search, update, and

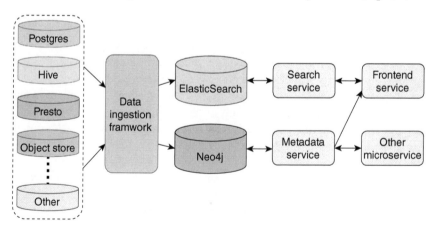

Figure 9.2 Amundsen architecture.

view metadata. The service makes use of search service and metadata service to power its user interface. The frontend service uses Flask and React to build the frontend.

Amundsen Search Service The search service provides a Restful API on top of ElasticSearch for other services to use, specifically frontend service. Search service is also a Flask application.

Amundsen Metadata Service The metadata service provides a Restful API to update table and column descriptions and table tags. The metadata service uses Neo4j or Apache Atlas for the persistence layer. There are a few advantages of using Apache Atlas as a persistence layer for Amundsen. Apache Atlas integrates with other Big Data systems. Moreover, it uses Apache Ranger for access control. If Apache Atlas is in the Big Data platform ecosystem, it is probably better to leverage it over Neo4j.

Amundsen Databuilder The Databuilder is a data ingestion framework. We can use the framework as a standalone script or inside an orchestration system like Airflow. A Databuilder job consists of extraction of records from the source, transform records, if necessary, and load records into the sink. The Databuilder component has concepts such as extractor, transformer, and loader to successfully transfer metadata. A job is the parent component of a task where each task consists of an extractor, transformer, and loader. Extractors pull data from the source, transformers transform the record to the target type, and the loader pushes data into the sink. Amundsen has several types of extractors, such as Hive, Cassandra, Postgres, and so forth.

9.3.3 Apache Atlas

Apache Atlas is a data governance software designed to collect, organize, and store metadata. Atlas provides foundational governance services to meet data discovery, metadata management, and compliance requirements. Atlas manages a central metadata catalog for data assets through the organization. It enables data engineers, data analysts, and data engineers to collaborate and discover metadata (Atlas, 2020). Out of the box, Atlas provides the following metadata capabilities.

- Atlas offers a type system. It can ingest existing types as well as creating new types. It provides a Rest API to access and manage types.
- Atlas helps classifying data. We can classify data as PII, SENSITIVE, etc. Data assets can have multiple classifications that help in security, compliance, and discovery.
- Atlas has a user interface to visualize data lineage. It provides a Rest API to control data lineage.

- Atlas implements comprehensive searching capabilities to discover and view metadata. It provides both Rest API and DSL (domain-specific language) to make advanced searches.
- Atlas integrates with Apache Ranger to establish access control.

9.3.3.1 Searching

Atlas provides two types of searching mechanisms basic and advanced search. Basic search allows users to query metadata with the type name, e.g. *Hive_column*. It also can filter based on a tag, classification, and entity attributes. The user interface helps to construct complex searching criteria with and/or expressions. All of the filterings get converted into an expressive Rest API call. Advanced search uses DSL to emulate SQL-like interface for additional expressiveness. With this additional abstraction, users do not have to know details of the underlying storage mechanism in Atlas. A sample DSL query is as follows:

```
from Table where createTime > '2018-12-31' AND name LIKE '*_zul'
```

9.3.3.2 Glossary

Apache Atlas offers a glossary where we can define business terms and link terms to each other. The terms in the glossary can be later mapped to databases, tables, and columns. The glossary helps to understand the purpose of tables and columns by providing additional business context. Moreover, terms help to search and discover by vocabulary, which is familiar to the users. Atlas offers a rich user interface for glossary functionality. We can add/update terms through the user interface as well as Rest API.

Atlas implements the glossary by building capabilities around the term, term category, and term hierarchy. We can add and update terms to the glossary at any time. A term belongs to one glossary. Its lifetime is bound to the lifetime of the glossary. We can assign terms into categories to give a more effective meaning. For instance, we can have a finance category, where we can have financial terms such as liability or expense. We can also create a hierarchy of terms where we can have broader terms and finer terms. For instance, we can have a term loan, where its parent is a liability.

9.3.3.3 Type System

Apache Atlas provides functionality to manage metadata models. The models consist of definitions called types. Atlas provides a type system where we can define and manage types for data modeling. All metadata models in Atlas are defined in the type system, such as Hive tables. Types define how a model is stored and represented in Atlas. A type represents one or many attributes that define a data model or entity. We can think of types as table schema or data classes. Type system in Atlas looks like as follows in Figure 9.3.

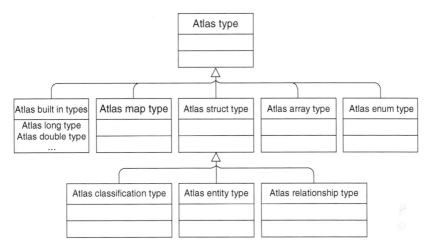

Figure 9.3   Atlas types.

Atlas has a base type called Atlas type. Every other type extends Atlas type. Map type, array type, or primitive types extend Atlas type. Atlas represents complex types such as classification, relation, and entity types under struct type. Struct types are not initialized directly. Subtypes of struct types such as entity type get initialized. Entity types have attributes where we can define relevant information for an entity like an owner or description. All Atlas types are registered to the Atlas type registry.

Atlas handles type information automatically with the systems it integrates, such as Hive. Atlas comes with a couple of predefined entity types that can be system-specific. For instance, Atlas has *Hive_db*, *Hive_table*, *Hive_column*, etc., for Hive. We can also add entities that are different than tables, such as HDFS path, processes. Atlas type system is generic enough to add any kind of information to the repository.

9.3.3.4   Lineage

Atlas allows linking of one entity to another through lineage. Atlas clients can send lineage information to Atlas. Once the lineage information arrives in Atlas, it can then be visualized through the user interface. The lineage population can be part of the workflow orchestration. For instance, Airflow integrates with Atlas and sends lineage information to Atlas. Once lineage information are populated, Atlas can show the full journey of the data from origin to the target aggregation.

Furthermore, Atlas offers classification propagation on the lineage path. If one or more entity gets a classification such as PII, Atlas propagates classification information to all impacted entities in the lineage path. For example, if we create an external table from HDFS and mark the address column as PII information.

The later tables would have PII classification. On the flip side, we can also stop propagation by marking it not to propagate.

9.3.3.5 Notifications

Atlas captures metadata changes and publishes them to Kafka topic with name *ATLAS_ENTITIES*. Atlas sends notifications for entity updates such as additions and deletions. Moreover, it sends notifications for classification addition, deletion, and updates. We can listen to these notifications on a client and send notifications to interested parties. Nevertheless, there is not a very good way to listen to only certain metadata updates.

9.3.3.6 Bridges and Hooks

Atlas integrates with external systems through bridges and hooks. Bridges let external systems asynchronously update metadata information in Atlas. Bridges provide a one-time update to the Atlas. This is generally used when the platform is initially loading. With a bridge, Atlas can load the current state of metadata. A hook is a listener for an external system, where it captures changes and notifies Atlas. Atlas comes with some hooks already. For instance, Hive, Storm, HBase, Sqoop, Kafka, and so forth. We can also build a custom hook. For the integrated systems, Atlas uses predefined types so that hooks can directly publish information.

If we would like to integrate a custom system, we can publish changes to a Kafka topic, *ATLAS_HOOK*, or calling API. To publish updates, we first have to register new types for the system. Once we register new types, we can then publish metadata updates to Atlas. Atlas captures all changes through hook notifications that have a notification type that can be type creation, type update, entity creation, and so on. Atlas receives these notifications, applies them to the registry and stores the result in the database.

9.3.3.7 Architecture Overview

Apache Atlas has a layered architecture where it interacts with external systems through Rest APIs and Kafka messages. Metadata sources publish the updates with hooks and bridges. Atlas powers the user interface through Rest API and integrates with Apache Ranger for policies on entities. At its core, Atlas has ingesters, exporters, a type system, and a graph engine. Atlas uses JanusGraph, an open-source distributed graph database to store metadata information. Janus-Graph uses HBase to store data but can be configured to use other backends. Atlas uses Apache Solr for indexing and searching (Figure 9.4).

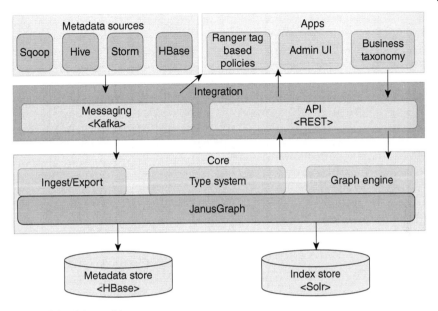

Figure 9.4 Atlas architecture.

Atlas keeps the metadata as an interconnected graph. Graph model gives the flexibility to add information and an efficient way to retrieve relationships between metadata nodes. Atlas handles translation of entities to Atlas type system, the persistence of graph model, and creation of indices for searching purposes through graph engine. Graph engine interacts with JanusGraph to retrieve and update metadata objects. Moreover, it interacts with Apache Solr for indexing.

10

Data Security

After reading this chapter, you should be able to:

- Learn infrastructure security patterns
- Understand data privacy and techniques
- Describe prevalent law enforcement on data
- Learn Big Data security solutions

Infrastructure security and data privacy are a must but a hurdle for any organization at the same time. Organizations want to take advantage of Big Data technologies while addressing security and privacy concerns. Big Data makes matters a bit more complicated as various data sources with different policies meet at the same data storage. Large volumes of data from partners, customers, subjects, and products meet in the same place. Big Data platforms are double-edged swords. The aim here is to ease the development of insights as easily as possible and to mitigate any vulnerabilities. While designing Big Data platforms, concerns about infrastructure security, data privacy, and addressing law enforcement arise.

10.1   Infrastructure Security

Securing Big Data platforms requires securing underlying infrastructure components, which mostly require distributed computing. Distributed computing means partial data per node but much more complexity. The more complex the systems are, the easier the systems to have security flaws and issues. Recently, distributed computing frameworks or storages that empowered most of Big Data infrastructures lacked proper security protocols. Distributed computing frameworks need to provide security during computation between nodes. Distributed storages need to provide access control management. Lastly, Big Data processing technologies might need to address common security vulnerabilities.

10.1.1 Computing

Big Data processing requires more than one machine. The abundance of machines results in multichannel communication and coordination among machines. Increased interaction and complexity lead to security problems of all sorts. The problem space includes machine-level security to network security. Active and passive security measures must be utilized to cater to security needs such as auditing, operating system, and network.

10.1.1.1 Auditing

Most systems generate audit logs to show "who" did "what" activity and "how" the system operated. Audit logs can reveal a complete picture of normal and abnormal events on a cluster. Audit logging should be enabled for Big Data systems and frameworks, which should be secured with proper permissions. From audit logs, information about the addition or deletion of nodes, state changes, and so on can be learned Gaddam (2015).

Audit logs aid in understanding potential security breaches, information misuse, or damage assessment. Audit logs can be stored in a search platform to enable the fast detection of abnormal activity and can be fed to a stream processor to alert on abnormal activity. For example, we can have audit logs for HDFS and YARN. These audit logs are sent to ElasticSearch and made available for system admins to check. These audit logs can be pumped to Apache Flink to detect any abnormal activity.

10.1.1.2 Operating System

The operating system comes into the picture in two ways: the first one is the choice of the operating system, and the second one is keeping the operating system up to date for security patches. Linux has different distributions, but Security-Enhanced Linux (SELinux) distribution provides additional security features. SELinux gives more control to administrators about who can access to the system, implements mandatory access control (MAC), and checks for allowed operations to limit the scope of damage to the machine. SELinux can be run in a permissive mode, where actions against the policy are allowed but logged for auditing purposes. Once an operating system of choice is installed, security updates should be observed. The process should be automated where possible, and machines should be patched as soon as an update gets released.

10.1.1.3 Network

Restricting access to the cluster nodes through access control lists) (ACLs) is one of the common solutions to limit unauthorized access to the cluster at the network level. ACL can be put to the top of the racks or virtual private networks to restrict access to the nodes other than authorized applications at the network. Moreover, firewall rules can be set to restrict access to only certain nodes in the cluster like Hadoop NameNodes. Lastly, proxy solutions can be employed to disable access to the nodes directly and provide access only through proxies.

10.1.2 Identity and Access Management

Identity and access management consists of two key components: the first one is the authentication process, and the second one is the authorization process. Both processes may be used to make sure access to data comes from allowed clients.

10.1.2.1 Authentication

The identification of applications and clients is the first step toward securing a Big Data environment. The identification happens through authentication processes, where the system checks if the clients have the identity of whom they claim to be. The authentication process generally involves a username and password verification of clients. Nevertheless, username/password authentication methods should be avoided when possible. For instance, authentication can be avoided in many places by leveraging a security assertion markup language (SAML) implementation. Once the identification process has completed, the system should require authorization mechanisms for finer control over resources.

10.1.2.2 Authorization

When the client tries to access an object or a resource, the Big Data system should allow access to objects pertaining to the client's permissions. The approval of access to objects happens through authorization mechanisms. The system can check whether the user has access to a certain object by comparing its role to that roles that an object or a resource has.

A role is a collection of access rights to one or more objects or resources in storage systems. With the help of roles, data can be managed by roles instead of users. Having roles instead of users can also avoid problems when a person leaves or joins the company. For instance, a person might have a Hive table. To delete or change a table, one needs administrator support. Moreover, roles can reflect organizational and team structure when it comes to accessing data. LDAP group membership can be used to enforce access control across all data.

Even if the use of roles in one or more systems is established, it is often hard to apply global permission management when organizations use many different storage systems with diverse processing engines. In such cases, a useful approach is to classify data in different storage systems through data discovery tools. Perhaps, audit the ownership information as well as roles assigned to certain object through discovery tools.

10.1.3 Data Transfers

Big Data systems often require consolidating data from different data sources. Moreover, summarized data might be pumped back into production databases for different purposes. For instance, we can calculate a rank for items and load item rank information daily or hourly back to the production database. When accessing to production databases, there are two common problems: securing access to the production resource and constraining the resources used by transfers.

The organization can keep production databases in a private network. Production databases might have firewalls or require credentials to access the resource. When accessing production databases, computing should be restricted to a certain group of nodes in the cluster. Furthermore, if the production resources require credentials, storing credentials should be avoided when possible. Instead, credential management software might be used if the organization supports one.

When accessing the production databases, access should be restricted to read-only roles when only transferring data out of the production database. When updating existing information in a certain table, the access should be restricted to that table with a role. Moreover, the resources for that role should be constrained when possible through queuing or bucketing whichever mechanism is available. It is a common scenario to have an overload on production databases because of data transfer. Ideally, data transfers should not bother production databases.

10.2 Data Privacy

Data privacy is concerned with the collection, sharing, and processing of personal data and requires systems to comply with the law and protect individuals' private or personal information while processing data. Many organizations might give up on implementing Big Data solutions because of the issues around data privacy. Although establishing data privacy is challenging, well-known techniques can be applied to overcome them. Encryption, anonymization, and perturbation methods can be used. Note that we, as designers of Big Data platforms, should put extra care for individuals' information. In the end, it is our responsibility to protect users from data breaches.

10.2.1 Data Encryption

Strong security requirements can necessitate data encryption. Data encryption may also be mandatory due to the nature of data in some organizations and provides a way to decode the information into a form such that it can only be decrypted with the correct key. Encoding and decoding information needs to be applied to both data at rest and data in transit to secure the environment thoroughly. The encryption strategy depends on the circumstances, for instance, self-hosted vs. cloud provider. Depending on the environment, encryption can take place in different layers, wherein each layer has its relative merits. Encryption layers will be discussed briefly.

10.2.1.1 File System Layer Encryption

File system layer encryption can scale easily across the cluster. It can provide high-performance encryption with an easy way to deploy. The encryption will become transparent to the applications running on the file system. Nevertheless, this might not be the best option since it does not give much choice to application or database running on top of the file system. Some applications might not need encryption at all and so do not need to pay tax for encryption. On the database side, users might want to encrypt certain columns in a table with their choice of encryption method. File system layer encryption is supported by HDFS and object storage solutions by cloud providers. For some domains such as banking or finance, the file system layer encryption might be a good choice since they might want to encrypt almost everything.

10.2.1.2 Database Layer Encryption

Database layer encryption brings flexibility and offers potential performance improvements for columns that do not need to be encrypted when the underlying storage is columnar. Most database vendors including the Hadoop ecosystem offer table or column-level encryption. Mostly, indexes or partition keys cannot be encrypted; it is less likely that we would need to encrypt them.

10.2.1.3 Transport Layer Encryption

Transport layer encryption uses secure sockets layer (SSL)/transport layer security (TLS) protocols to move data between client and server or servers. SSL/TLS protocols provide encrypting the packets transmitted between machines. With transport layer encryption, data in transit can be secured. Setting up the transport layer encryption involves setting up public/private keys and storing them securely a secret store. Transport layer encryption is a must-have when working with highly critical data. Most of the Big Data solutions provide encryption in transit out of the box.

10.2.1.4 Application Layer Encryption

Application layer encryption is a secure and flexible choice and has the ultimate authority to what to encrypt and gives a ton of choice to the developer. However, handling encryption in all applications requires a lot of work especially if the table or data source gets shared among different applications. Hence, application layer security is generally not very practical.

10.2.2 Data Anonymization

Data anonymization is a technique that removes personally identifiable information (PII) data from the original data so that the remaining data is anonymous. Anonymization is a critical technique to support the privacy of individuals while generating business value and needs to be applied before data gets published for further processing. The main anonymization operations are generalization, decomposition, replacement, suppression, and interference. Anonymization operations aim to achieve privacy-preserving methods such as k-anonymity, l-diversity, t-closeness, and differential privacy (Fang et al., 2017).

10.2.2.1 *k-Anonymity*

k-Anonymity is a privacy protection method that helps to hide the identity of individuals in a group of people. For each record in the data set, if there exists at least $(k - 1)$ records that have the same properties, then k-anonymity is achieved (Sweeney, 2002). For instance, imagine a data set where k equals to 30. At least 29 people who have the same age for each person should be found in the data set. Therefore, any individual cannot be identified by their age. The data can be processed further to generalize the age column into the broader category. For example, instead of age 23, $20 < \text{age} \le 25$ can be used, and this type of generalization can be applied to every record. For identifying columns but irrelevant to the context, the data can be potentially replaced partially or totally removed.

10.2.2.2 *l-Diversity*

k-Anonymization guarantees that any record is indistinguishable from at least $(k - 1)$ records concerning personally identifying attributes. Nevertheless, if all records share the same attribute, then sensitive information can be accessed by just knowing if an individual is part of the data set (Machanavajjhala et al., 2007). For instance, suppose a friend goes to a sports club and the sports club publishes anonymized data on attendees. If some people have the same attributes as the friend, each doing kung fu, then we can conclude that the friend is doing kung fu. l-Diversity makes sure the data has not only kung fu but also jiujitsu, therefore protecting the friend's privacy better. t-Closeness further refines l-diversity and gives better privacy protection (Li et al., 2007).

10.2.3 Data Perturbation

Data perturbation is a data privacy technique that adds noise to the data to address the privacy concerns of individuals. Data perturbation methods add random numerical values to private data to protect original values. Data perturbation aims to run aggregation jobs such as correlation while protecting individuals' private information (Wilson and Rosen, 2003). Additive perturbation can be used for some columns. For instance, a random number can be added to the age column of the user table in the database. Having small randomization would not change the aggregated result much, but it would protect the individual's privacy.

10.3 Law Enforcement

In recent years, tremendous developments have taken place when it comes to protecting individual's rights on data privacy from a legislative perspective. Companies have been trying to comply with new regulations, which guide companies in storing, sharing, and processing of data. While designing Big Data platforms, regulations should be taken as guidance, and individuals' privacy should be protected through conventional methods. In this section, some practical techniques will be discussed to manage PII data in Big Data environments. Moreover, we will touch on privacy regulations/acts and how privacy regulations/acts can be ensured in Big Data platforms.

10.3.1 PII

PII is any piece of information that identifies an individual. It can change from one jurisdiction to another. Nevertheless, it is any information that can distinguish one individual from another. Some of the examples of PII are the place of birth, social security number, age, name, etc. To summarize, PII is any information that can be linked to the individual such as education, medical, and employment information.

Protecting PII data is essential when data gets poured into common environments. In the early days of Hadoop, the security aspect of Big Data was not prevalent. Thus, many companies still lack security configuration. Transferring PII data to such environments should be strictly avoided, and methods to deal with PII data should be exercised. Such methods will be discussed to address PII concerns.

10.3.1.1 Identifying PII Tables/Columns
The first step of coping with PII is to define PII in the organization. Once PII is clearly defined, finding tables and columns that contain PII data can be started. For each table and column that contains PII, we should document why it is PII and categorize them as PII using the classification method. If data discovery tool is available, then it should be used to mark columns and tables PII.

10.3.1.2 Segregating Tables Containing PII
Once we classify what is PII and what is not, data transfers from different data sources can then be organized according to the classification. There are two methods we can apply with PII tables. Transferring PII data to common storage can be

avoided, where access is not restricted, or the table can be partially transferred. When transferring partially, a top view of PII tables can be achieved, which omits PII columns, and we can transfer the view instead of the PII table. Instead, a query can be run that selects only non-PII columns for a given table, and we can transfer the query result.

10.3.1.3 Protecting PII Tables via Access Control

If PII tables are needed, then access permissions with proper roles should be implemented. A schema such as pii_restricted can be created, and a role such as PII_RESTRICTED can be assigned to this schema. When PII data is moved to data lake or warehouse, pii_restricted schema would be used for PII data. If the aforementioned convention can be established through the organization, managing PII data would be much simpler. Another key point is to keep table names and columns the same between databases. Again, a data discovery tool can be leveraged to detect and monitor naming consistency between PII tables and columns.

10.3.1.4 Masking and Anonymizing PII Data

Transferring PII tables into the secure schema with restricted access permissions secures data. Nevertheless, regular users who want to use the table cannot access it. To address this problem, a top view of PII tables can be achieved, and PII columns can be masked with an asterisk, or some other value that shows it is masked. Another schema such as pii_masked can be created to add views under this schema. Moreover, anonymization operations can also be used such as a generalization to remove PII information. An example view is as follows:

```
CREATE OR REPLACE VIEW pii_masked.customer_vw AS
    SELECT customer_id,
        '*' AS customer_phone,
        '*' AS customer_name,
        CASE
            WHEN credit < 2500 THEN 'low'
            WHEN credit > 2500 AND credit < 5000 THEN 'middle'
            ELSE 'high' END
        AS credit_group,
        CASE
            WHEN age <= 35 THEN 'young'
            WHEN age > 35 AND age <= 55 THEN 'middle_aged'
            ELSE 'older' END
        AS age_group
    FROM pii_restricted.customer;
```

10.3.2   Privacy Regulations/Acts

New legislations are implemented recently to empower individuals over their data and how it is stored, used, and processed. The legislations only apply to a region or a country, but many countries have been implementing legislation to protect the privacy of individuals. General Data Protection Regulation (GDPR) and California Consumer Privacy Act (CCPA) are prominent legislations. GDPR is focused on creating a legal framework across the European Union that expects companies to pertain privacy regulations by default and expects organizations to approach data protection by design (Politou et al., 2018). On the other hand, CCPA is focused on Californian consumers and provides transparency and control over how organizations collect and use their data.

Under GDPR and CCPA, individuals get many rights on their data, but specifically, the focus will be on collecting data and erasing data.

10.3.2.1   Collecting Data

Cookies are an important way of tracking users throughout their journey on the website. The cookie can be logged to access logs or streams and be used to create a better overall experience for the user. With regulations in place, one might opt-out of cookies on the website, where at that point the user cannot be tracked. However, the user can still be tracked through their IP and some other relevant information. From the data platform perspective, regulations change how much data can be passed down to our data platform. Thus, not much is needed to be done when it comes to the collection of data.

10.3.2.2   Erasing Data

Deleting data is a hard and quite cumbersome process in Big Data platforms. A quite big portion of Big Data storage systems does not allow deleting individual rows. However, there are levels of implementing data deletion such as new data generation, removing existing data, and anonymizing data.

If the user gets deleted from the system, practically marked as deleted, protection mechanisms should be in place such that any new data for the user will not be processed. For most ETL pipelines, the member, customer, or user data pumped into the data lake in daily or hourly cadence. The more frequent the reload operation can be done, the better we can decide which records to avoid processing. Thus, the key is the frequency of information updated in data pipelines.

If the data can be kept anonymized or masked, then keeping them is less harmful. Our system should still be designed to delete data as early as possible; however, the threat surface is much smaller. Pouring personal data into a common data lake can be avoided, and a technology can be used such as Presto to query everything live from a read-only replica of the production database. With this approach, processing new data for a removed user can be avoided.

Last but not the least, any data that get written to a different set of storages should have a retention policy. Thus, after the removal of the user from the system, the platform should forget about the user at some point. Having a default retention policy can also remove the problem from backups. If some part of data get deleted in the original storage system due to retention, then a daily backup generated will not include deleted records.

10.4 Data Security Tools

Big Data platforms entail combining many systems and applying security best practices across the board. When it comes to tooling, many options are available from cloud providers to open-source software. Each toolset offers different security mechanisms to protect platforms from potential data leaks, integrity problems, access control issues, etc. In this section, some of the open-source security tools for the Hadoop ecosystem will be discussed to see how they help with common security concerns.

10.4.1 Apache Ranger

Apache Ranger provides a dependable security solution for a Hadoop ecosystem. Ranger supports the capability to establish, manage, and maintain security policies for all components in the Hadoop ecosystem and can enforce policies on files, databases, tables, and columns. Policies can be set per user or group. Ranger allows Hadoop administers to audit Hadoop components and tracks all access requests. Moreover, it can provide reports to summarize policies. The more policies Ranger has, the more difficult it gets to manage them. Ranger gives visibility over Hadoop components (Ranger, 2020).

10.4.1.1 Ranger Policies
Ranger offers two types of policies for the management of components: resource-based and tag-based policies with conventional access conditions.

Resource-based Policies A resource-based policy allows an administrator to grant permissions to users and groups on one or more objects on the target services. For instance, a resource-based policy can be applied on an HDFS path.

Tag-based Policies A tag-based policy allows an administrator to grant permissions to tags. Ranger tag service is responsible for finding tags for a requested resource.

Ranger caches these tags assigned to a resource for faster lookup. After finding tags, it evaluates tag policies on the resource. If one or more tags implicate to deny accessing the resource based on the accessor, then Ranger denies access to the resource. If the requested resource has no tags, Ranger defaults to resource-based policy to grant permission on the resource.

Ranger allows conditions for evaluation of tag policies. Ranger prepares a request context where many details about the resource and accessor are available. For instance, Ranger can use a client's IP address to deny access to a resource. The request context can be customized with context enrichers by extending *RangerAbstractContextEnricher*. Furthermore, tags have an expiry date. If the tag expires, the policy would not be applied to the resource.

Access Conditions   Ranger provides Allow, Deny, and Exclude conditions as follows:

- Allow
- Deny
- Exclude from Allow
- Exclude from Deny

Ranger access conditions give enough flexibility to define access conditions on a given resource. For instance, many access conditions are defined in the data database called revenue. Anyone from the executive team might have access to such resources. Thus, we can Allow the *executive_team* to access the resource. If data engineers are working on a revenue database, we can Allow them with their group *revenue_engineers*. Contractors can be excluded with Exclude from Allow conditions. Ranger evaluates policies in a particular order to guarantee expected outcomes. The Ranger policy decision flow is depicted in Figure 10.1.

10.4.1.2   Managed Components

Apache Ranger provides a centralized security framework to manage access for YARN, HDFS, Hive, HBase, Storm, Knox, Kafka, and NiFi. In this part, some of these components will be discussed to see how Ranger helps access management for them. Note that Ranger needs an authentication mechanism to restrict access. A common authentication method can be set up such as Kerberos or LDAP to control access to resources.

YARN   Ranger allows us to control the usage of YARN queues. A queue policy can be assigned to a user and a group. Having a policy on the YARN queue is important as the entire cluster can be drained down by one rogue query.

HDFS   For HDFS, Ranger checks if there exists a policy for a given directory or file. If a policy exists, access is granted to the user. If a policy does not exist, then

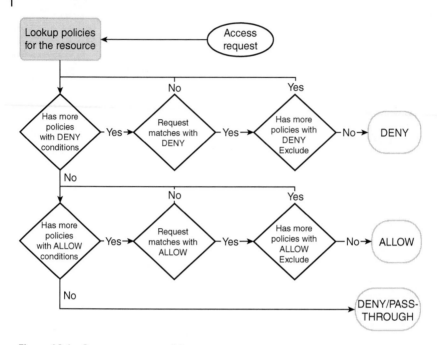

Figure 10.1 Ranger access conditions.

Ranger would fall back to the native permissions model (POSIX). Nevertheless, it is important to set Ranger policies on data folders for applications such as Hive. Ranger allows defining policy by the path with user or group. The policy can apply for one or more paths.

Hive Ranger provides fine-grained access control over Hive and provides a table or column level access permissions. In Ranger, you can select database, table, and column names to apply access control. Users or groups can be assigned for any policy. Ranger can apply dynamic resource-based column masking to protect sensitive data in Hive. Policies that mask or anonymize sensitive data columns can be set, such as PII dynamically for Hive. Lastly, Ranger can apply row-level filtering based on the groups. It can automatically filter rows based on the user or group. An expression to users or groups can be entered, and Hive will output results by applying these filters based on the Ranger policy.

Storm Ranger provides a similar access control mechanism for Apache Storm. An access level can be assigned per topology. Users and groups can be assigned to the policy we have for Storm. Ranger provides access controls like submitting or killing a topology.

Kafka Ranger provides access control over Kafka. Ranger can be used to restrict consumers and producers on topics. Users or groups can be added per Kafka topic. If Kafka is used for user events such as creation/updates and passing down PII data, Ranger integration can be critical. User information should not be leaked to every consumer that asks for it.

10.4.1.3 Architecture Overview

Apache Ranger has a simple and extensible architecture. Each component in the architecture is easy to extend or replace. Integration and plugins do not require code change in Ranger. On a high level, Ranger consists of servers, plugins, and integration to external systems (Figure 10.2).

Ranger Admin Server Ranger admin server is a central component to administering ranger policies. The admin server provides a user interface where policies can be defined, managed, and controlled. Moreover, it allows administrators to generate reports. Apart from user interface functionality, it provides an API for external systems to integrate with it.

Ranger Policy Server Ranger policy server integrates with plugins, tag synchronization, and user/group synchronization. It is a bridge between policies, users, and plugins. Ranger plugins retrieve policy information through the ranger policy server periodically. To avoid roundtrip, plugins cache policies locally. This is

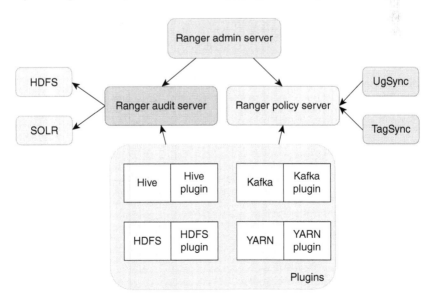

Figure 10.2   Ranger architecture.

necessary for components as it would not be possible to enforce access permissions without knowing policies. The policy server queries the component regularly for objects that live on the component. Ranger needs object information for auto-completion and dropdowns when defining policies on components. Nevertheless, policy server to component plugins is not essential as policies can still be defined.

Ranger Audit Server Ranger audit server writes audit events to HDFS and Solr. Ranger uses Solr for displaying audit data. Audit server uses HDFS for backup purposes only. It also supports audit log summarization. Similar logs get aggregated into single audit entries.

Ranger User/Group Sync Ranger user/group synchronization integrates with Kerberos, LDAP, and Unix to retrieve user/group information. User/group information gets stored in the database and is used in defining policy definitions.

Ranger Tag Sync Ranger tag synchronization integrates with Apache Atlas for retrieval of tags. Ranger tag synchronization is event-based. Changes in different components such as Hive can create an event to Kafka topic *ATLAS_HOOK*, and then, Atlas will pick up the changes. Changes in Atlas will result in publishing an event to Kafka topic *ATLAS_ENTITIES*. Ranger Tag synchronization consumes this topic and picks up the changes.

Ranger Plugin Ranger plugins are lightweight Java programs that are installed on the target component such as Hive or HDFS. Plugins pull policies regularly and cache them locally. They act as an authorization gateway and help to authorize requests coming from different clients. Ranger allows plugins to be easily installed without change to the Ranger itself.

10.4.2 Apache Sentry

Apache Sentry brings role-based authorization to Hadoop components. Sentry provides enforcing access policies on data by integrating with LDAP, AD, etc.; is a pluggable authorization mechanism for Hadoop components; allows authorization for Hadoop components such as Hive, HDFS, Solr, and Impala out of the box; and gives the ability to define authorization rules on different data. Sentry's modular nature allows it to integrate with any data source (Sentry, 2018).

10.4.2.1 Managed Components
Apache Sentry integrates with many Hadoop components. Apache Sentry server plays a central role for Hadoop components. It stores authorization metadata and provides APIs for tools to retrieve and modify this metadata securely.

The authorization happens in the policy engine that runs in the Hadoop components. Sentry can integrate with major Hadoop components like Hive and HDFS.

Hive When a user submits a new query to Hive, Hive will identify the user and parse the query to see which tables the user wants to access. Hive then asks Sentry plugin to validate the access request. The plugin will retrieve the user's privileges related to the tables in the query, and the policy engine will determine if the user has access to the tables in the query.

HDFS HDFS plugin complements the authorization mechanism that happens in Hive. The purpose of HDFS integration is to expand the Hive authorization checks to any other Hadoop components such as Pig, MapReduce, or Spark. Sentry plugin does not replace standard rules. Any data path that is not in Hive metastore works with existing ACLs. HDFS NameNode integrates with Sentry plugin and authorizes access requests through this plugin. The plugin periodically pulls new authorization rules and caches them.

10.4.2.2 Architecture Overview

Apache Sentry employs a straightforward architecture and integrates with Hadoop components with plugins. It consists of a policy server, policy metadata store, plugins, integration to authorization, and group mapping.

Sentry Server Apache Sentry server handles the authorization and provides an interface to securely retrieve and manipulate the metadata. It saves policy metadata to its data store through Sentry store. Sentry store provides data access for sentry data and access privilege manipulation.

Sentry Plugin Sentry plugin works inside a Hadoop component. It consists of interfaces to manipulate authorization and a policy engine. The plugin interfaces help to manipulate authorization inside Sentry server. The policy engine evaluates access requests that a component receives from clients, and it uses the authorization metadata retrieved from the server to grant permission on a requested resource. The plugins try to synchronize with Sentry server by polling for updates regularly. Plugins cache policy information locally to decrease synchronization efforts (Figure 10.3).

User/Group Mapping Sentry uses authentication systems such as Kerberos or LDAP to identify the user. It retrieves the group mapping from the underlying system like LDAP. Thus, it has the same group mapping mechanisms with other Hadoop components.

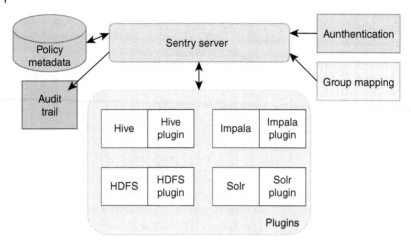

Figure 10.3 Sentry architecture.

10.4.3 Apache Knox

The Apache Knox is a gateway to the Hadoop ecosystem. Knox controls authentication and access for services residing in one or more Hadoop clusters and simplifies access to Hadoop services for both administrators and users by implementing a single endpoint. Knox implementation employs a pluggable design, where it enforces different policies. Knox policies range from authentication, authorization, audit, dispatch, host mapping, and content rewrite rules. Knox enforces these policies with a chain of providers defined in the topology deployment descriptor for every Hadoop cluster (Figure 10.4).

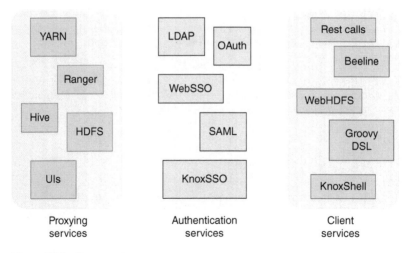

Figure 10.4 Knox services.

Knox introduces a new cluster with a topology deployment descriptor. The deployment descriptor configures each cluster, and clusters get its collection of REST APIs represented by a single cluster-specific application context path. The topology deployment descriptor gives the layout of the cluster for purposes of routing and translation between client-facing URLs and cluster internals. Knox gateway exposes a single URL that aggregates all Hadoop services. Therefore, it limits network endpoints to reach the Hadoop cluster and hides internal Hadoop cluster topology.

Knox gateway provides perimeter security where a secured boundary is present between internal and external networks. Perimeter security provides strong authentication to allow controlled access to Big Data services. Accordingly, Knox Gateway provides authentication and token verification and enables authentication integration with identity management systems for both enterprise and cloud solutions. Knox extends authentication capabilities with service-level authorization (Knox, 2020).

10.4.3.1 Authentication and Authorization

Knox uses two types of providers to identify the caller's identity. The first one is authentication providers that accept caller credentials and validates against a service. The second identity provider is federation providers that validate a token that has been issued to the caller by a trusted identity provider. Knox comes with an authentication provider that uses http basic to accept caller credentials and relies on LDAP to validate the caller's identity. Apart from the basic authentication, Knox uses comprehensive federation methods.

When the organization uses a federated authentication mechanism to limit the number of entities that validate user identity, Knox can federate the authenticated identity from an external authentication event. Knox can use Pac4J for the federation. Pac4J supports numerous authentication mechanisms such as SAML, CAS, and Open ID. Moreover, Knox offers a service called Knox SSO that supports integration with single sign-on (SSO) and provides an authentication token for the caller. Knox SSO hides the actual identity provider integration from applications; therefore, applications are only aware of Knox cookie. Knox also supports authorization over accessing services.

Knox provides authorization rules based on the user identity context. User mapping rules and authentication provider determine the identity context. Authorization providers use identity context and grant access to a resource based on the policies. Knox comes with ACL-based authorization where it can support authorization mechanisms based on user, user groups, IP address, etc. ACLs protect resources on service levels. For instance, we might have a Hadoop cluster that contains sensitive information. Access to this cluster can be restricted with ACLs.

10.4.3.2 Supported Hadoop Services

Apache Knox integrates with many Hadoop components. Some of the notable components are HDFS, Hive, Livy, Ranger, YARN, etc. On the base level, Knox

Table 10.1 Gateway to component mapping.

Gateway	https://gw-host:gw-port/gw-path/cluster-name/component-name[a]
Component	http://component-host:component-port/component-name

a) gw is shorthand of gateway.

introduces a mapping between URL presented to the outside of the world and actual web API. Some of the components might expose information about cluster internals through URLs, and the gateway rewrites these URLs to ensure internal cluster details are protected. Table 10.1 shows the URL mapping.

Knox comes with http basic authentication. With the following curl command, we can list files in an HDFS directory.

```
curl -i -k -u dbdp:dbdp-password -X GET \
    'https://localhost:8443/gateway/dbdp/webhdfs/v1/?op=LISTSTATUS'
```

Or we can create an ElasticSearch index with the following curl requests.

```
curl -i -k -u dbdp:dbdp-password \
    -H "Content-Type: application/json" \
    -X PUT "https://localhost:8443/gateway/dbdp/elasticsearch/big-data" \
    -d '{
            "settings" : {
               "index" : {
                   "number_of_shards" : 11,
                   "number_of_replicas" : 3
                }
            }
        }'
```

Hortonworks Sandbox is a good environment to get started with Knox gateway. Spending some time playing with it is recommended. The sandbox is available in various formats.

10.4.3.3 Client Services

Hadoop ecosystem suffers from a lack of software development kit (SDK) or REST Client libraries. Knox addresses this problem by providing SDK for Knox interactions. Knox provides a domain-specific language (DSL) solution to interact with Hadoop services proxied by Knox. Knox also offers an interactive shell environment that combines a groovy shell with the Knox SDK classes. Let us see an example for DSL.

```
gateway = "https://localhost:8443/gateway/dbdp"
password = "dbdp-password"
basePath = "/dbdp/examples/"
```

```
file = "dbdp.data.tar.gz"
statusDir = "/dbdp/status/"
hiveQuery = "/dbdp/hive/example.hql"

session = Hadoop.login(gateway, username, password)
//Replace HDFS file
Hdfs.rm(session).file(basePath + file).recursive().now()
Hdfs.put(session).file(file).to(basePath + file).now()
//Run hive sql
Job.submitHive(session).file(hiveQuery)
    .statusDir(statusDir).now()
//Run a sqoop job
Job.submitSqoop(session)
    .command("import --connect jdbc:mysql://localhost:3306/dbdp...")
    .statusDir(statusDir).now()
session.shutdown()
exit
```

We can also connect to Hive over the JDBC connection. Knox rewrites application URLs, but the application user interface can still be reached. If a job to YARN is submitted, it would be available through a web interface. The history server and others can be seen.

10.4.3.4 Architecture Overview

Apache Knox gateway is an implementation on top of the embedded Jetty web server. The gateway finds the request filter chain for a given URL. The filter chain then processes the request. The gateway offers a flexible mechanism to add a group of filter chains to secure access to Hadoop services. The gateway has two mechanisms to extend: service and provider. The service helps to add new http endpoints. For instance, Knox uses a service to plug-in HiveServer2. On the other hand, the provider allows adding new functionality across all services. For example, authentication happens through a provider. Providers can also expose APIs that services can use (Figure 10.5).

Knox loads service and provider configurations through a topology descriptor file. In the deployment phase, Knox converts the topology descriptor file into an executable (JEE WAR). It loads topology from conf/topologies and goes through each service and provider defined in the topology and processes them. Once the executable is ready, Knox uses internal container API to load executable dynamically. In the runtime phase, Knox uses the executable to process incoming requests from clients. When a request arrives, Knox finds an appropriate filter chain (Servlet Filter) from a map of URLs. A request enters each filter in the chain. At the last chain, Knox dispatches a request to actual service. The response goes through various wrappers and finally arrives at the client.

Knox runs as a server or cluster of servers that provides an entry point to one or more clusters. Knox servers are stateless; therefore, scaling the Knox gateway

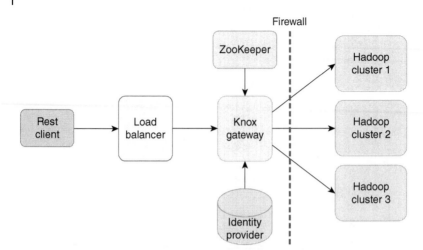

Figure 10.5 Knox architecture.

is reasonably straightforward. More Knox servers can be simply added horizontally to scale the gateway. Each gateway instance would have the same configuration and topologies. With the help of a load balancer, traffic can be distributed to multiple gateway instances. Furthermore, ZooKeeper can be used for topologies and configuration. Gateway instances can monitor ZooKeeper instead of local directories. Updates to provider configurations and descriptors in ZooKeeper will result in redeployments. Having ZooKeeper integration provides a more flexible deployment.

10.4.3.5   Audit
Knox gateway can log what actions were taken by whom. By looking at the logs, suspicious behaviors can be determined. The logging goes through the log4j framework but can be extended for other use cases. A typical solution might include ElasticSearch to quickly search for suspicious activity.

11

Putting All Together

After reading this chapter, you should be able to

- Compare Big Data platform design alternatives
- Review Big Data systems and tools
- Discuss various Big Data challenges

The previous chapters have discussed many Big Data platforms' components and modern Big Data platform technology requirements. The technologies used in different settings and the tools and systems addressing Big Data platform demands are explored. This chapter discusses the challenges while designing Big Data platforms.

11.1 Platforms

Designing and building a Big Data platform requires decisions. Working solutions are available and often used by organizations to deal with Big Data. However, these solutions might be incomplete, incoherent, and hard to maintain. An overall vision for the platform would help organizations. Nevertheless, shifting from systems to platform requires effort and dedication from the executive management and other parties involved.

Big Data platforms focus on infrastructure. One of the current key factors is the deployment model, which includes in-house solutions that require additional maintenance costs due to the provisioning of resources and talent. On the other hand, an infrastructure created by a cloud provider can be used to build the platform. The last alternative is to employ a hybrid infrastructure where some resources and systems are available from the cloud provider. The rest of the infrastructure is an in-house solution.

Designing Big Data Platforms: How to Use, Deploy, and Maintain Big Data Systems,
First Edition. Yusuf Aytas.
© 2021 John Wiley & Sons, Inc. Published 2021 by John Wiley & Sons, Inc.

The key element is the long-term cost and agility of the solution adopted. If a cloud provider is chosen and later becomes overly costly, it was a bad decision. On the other hand, if an in-house solution was chosen, keeping up with new technology, e.g. upgrades or maintenance, the system would not be agile enough. Hence, fundamental infrastructure decisions should be made under scrutiny as the systems' current state and infrastructure play an important role in decisions. In this section, the comparison of different infrastructure choices for Big Data platforms is discussed.

11.1.1 In-house Solutions

An in-house solution requires provisioning, upgrade, maintenance, and support of the software stack regardless of where it is deployed. In-house solution can be employed on top of on-premise hardware or managed resources by cloud providers. When utilizing an in-house solution, there is full control over the software. Depending on the needs, the software can be patched or adapted to the environment. The flexibility comes with a cost. Any system that a company maintains needs administration. The administration level depends on the software, but the administration cost is always part of the problem. Moreover, the level of administration depends on the infrastructure that software runs on. If it is an on-premise infrastructure, administrators need to worry about hardware failures as well as software problems.

11.1.1.1 Cloud Provisioning

Cloud provisioning of hardware makes it easy to scale by adding more nodes to the infrastructure through cloud provider API or user interface. It is also possible to create cloud provider-specific images and deploy them automatically. The software updates or maintenance of the overall system is relatively easier as it does not cost anything to remove virtual hardware. If one or more nodes in the overall system get sick for any reason, the administrator can replace the sick node with a new virtual node. Most Big Data solutions are horizontally scalable. Adding/removing nodes is not burdensome.

Although cloud-based hardware provisioning is comfortable, the software still needs maintenance and support. A dedicated team of administrators to support the rest of the company is needed. There would be onboarding processes for new team members and time to ramp-up with the infrastructure built. What's more, cloud-based hardware provisioning is costlier than actual hardware. The ability to easily provision new nodes comes with an overall cost on the budget because virtual hardware is not cheap. It often can be an order of magnitude expensive than traditional hardware.

11.1.1.2 On-premise Provisioning

On-premise hardware provisioning requires a different level of expertise in both technical and nontechnical abilities. It requires knowledge of Big Data systems software and hardware configuration about servers, racks, and switches. It also entails new roles, such as data center technicians. Ultimately, on-premise provisioning often needs an upfront investment and ongoing planning for growth.

The maintenance of bare metal hardware involves different processes and automation. There is hardware provisioning software such as ansible, chef, and terraform to define infrastructure as a code. Moreover, hardware provisioning requires experience in hardware and negotiation with hardware providers. Hardware warranty, support, and the price are all part of the maintenance process that the company needs to support.

Hardware provisioning also requires some physical work. One or more data center technicians are needed to put together new racks, switches, and servers. In the case of hardware problems in various components, they have to be engaged. Moreover, hardware problems might need to be addressed immediately. Having such requirements end up having on-call duties for technicians and administrators.

If the company does not have an existing on-premise infrastructure, it is really hard to invest in such an area just for Big Data platforms. If the company has to maintain its infrastructure due to scale or compliance, it is easier to convince the executive team to sponsor on-premise Big Data systems. Even after the initial investment, the team has to plan for the growth and maintenance costs such as hardware failures. The growth can be particularly estimated due to trends unless the company sees huge spikes in usage.

Although it seems quite painful to have an on-premise infrastructure for Big Data platforms, it can also be very cost-efficient. Once the hardware is bought, the rest of the expense is pretty much electricity. If the company already has expertise in maintaining infrastructure, it would not be a huge undertaking to provision a new cluster of machines. Furthermore, the performance of bare metal servers can be far better than virtual hardware. With on-premise hardware, the company also gets a chance to upgrade physical hardware such as disks while maintaining some parts.

Bare metal instances are also available from cloud providers. Nevertheless, they are mostly about renting the hardware, not purchasing it. Purchasing actual hardware and tuning it keeps the cost down. Otherwise, it would not make a big difference and might not be a good choice given that there are additional maintenance problems.

11.1.2 Cloud Providers

Cloud providers implement Big Data systems on top of open-source software solutions. They offer similar systems with pre-configured parameters optimized to run on their infrastructure. Additionally, they provide data storage facilities, such as object storage and data warehousing solutions. Most cloud solutions integrate very well within the boundaries of a cloud provider. Integration points can exist either through workflow solutions or some scripting that runs on serverless computing. There are many reasons why a cloud provider might be a good fit for Big Data platforms, such as affordability, time-to-market, ongoing cost, talent, and scalability.

The upfront investment cost for Big Data systems might be a big problem for smaller organizations. Investing little money on infrastructure for a small but fast-growing organization might become a hurdle. Moreover, there is no guarantee that the overall platform would work for the organization. At times, it might be wiser to buy a service rather than implementing the solution itself. In such cases, bashing the existing platform and outsourcing Big Data needs to a provider might satisfy the needs.

Cloud providers make it very easy to build a Big Data platform rapidly. Many recipes are available to build data storage systems, data processing, data discovery, and data science. These systems integrate seamlessly into a cloud provider. From data transfers to data processing, each cog is ready to implement the overall platform. Moreover, the iterative development of system components is possible. Therefore, building an minimum viable product (MVP) is easier. Once the MVP is built, scaling the platform is more natural as it is very probable that the systems underneath the platform can scale horizontally by adding nodes.

Cloud providers also charge the company a steady base for the expenses of the infrastructure. There is no upfront payment. At any point, the infrastructure can be destroyed, and there is no need to use it all day long. Spark application can be used to read data from the object store and save the result to a data warehouse. Hence, there is no reason to have a permanent provision of resources for such tasks, and the organization can save money.

Although cloud providers offer excellent benefits, they still have downsides like vendor lock-in and outages.

11.1.2.1 Vendor Lock-in

Cloud providers offer many solutions not only for Big Data but for many other areas. An organization that depends on the cloud provider can seamlessly use solutions within the cloud provider. Nevertheless, there is no standard for many solutions. Interoperability or portability between cloud vendors is not great either. Once a Big Data platform gets built-in one cloud provider using their offering, it becomes increasingly hard to move to another provider and requires many technical and nontechnical challenges.

An organization can try to stay away from solutions that require tight integration with a cloud provider. Nonetheless, this strategy can limit the number of solutions available and might require additional software components to be built.

A good solution might be an abstraction layer on top of cloud providers. Most cloud providers offer object stores with similar interfaces. A standard interface can be implemented for all common operations and integrated with the cloud provider. Hence, the cloud provider in the infrastructure would become a configuration. If the same strategy is applied to all components, the vendor lock-in problem can disappear.

11.1.2.2 Outages
Cloud providers promise high uptime guarantees. In most cases, they even beat the uptime guarantees. However, the cloud provider can go offline for a service or a couple of services for a region. If the Big Data platform hosts critical components such as fraud detection, it could become a huge issue. Depending on the risks and budget, platform components can be designed to live in multiple cloud providers. Nevertheless, such systems would not be cheap in development and maintenance.

11.1.3 Hybrid Solutions
Hybrid solutions were briefly discussed in the Big Data storage chapter. On-demand computing was the main approach. Cloud providers support on-demand computing by creating and destroying a computing cluster. The on-demand computing model helps to finish jobs quicker with a larger cluster and costs only when the job runs. The idea for a hybrid platform resembles the same model. Experimental setups can belong to cloud providers due to ease of development.

On-demand computing enables the processing of offline tasks such as large daily ETLs or model training. Upon receiving results for these tasks, the output can be streamed down to an on-premise cluster. The tricky part is to build tooling to support computing both on-premise and cloud. Ideally, it should be invisible to data analysts, data engineers, and data scientists how the computing gets done. The only thing that matters is the output of the computing. Enabling on-demand computing requires a setup where on-premise and cloud providers can share data. For instance, an object store might share data between an on-premise cluster and cloud solutions like data warehouses and data marts. Nevertheless, some of the needs might not be met very well with this model, such as stream processing.

Experimental projects can always leverage cloud provider solutions as a head start. Once the project gets concrete, maybe utilizing an on-premise cluster or investing in a particular solution might become an idea. Cloud providers enable building solutions without taking much risk other than human resources. Nevertheless, it might be harder to implement the same infrastructure on-premise cluster and stabilize it. Cloud providers offer better stability simply because they serve many customers and have to address many edge cases.

Although hybrid solutions provide cloud solutions access, it still has problems such as vendor lock-in and downtime. Another alternative is to use containerized software to build solutions on top of Kubernetes.

11.1.3.1 Kubernetes

Kubernetes is a powerful open-source container orchestration system. At a high level, Kubernetes is a system for running and orchestrating containerized applications over a cluster of machines. It manages the complete life cycle of containerized applications and has become the industry standard available to all cloud providers. This book has referred to containerized applications and Kubernetes several times. Kubernetes is becoming a popular system where Big Data applications can run easily. The advantage of running Big Data applications on Kubernetes comes from interoperability. For example, the same Spark job can run on-premise Kubernetes or cloud providers.

Due to the rapid adoption of Kubernetes in many areas, the new version of Big Data solutions come with direct support to Kubernetes. Applications like Spark and Airflow support Kubernetes natively. Moreover, most data science applications come with direct integration to Kubernetes for training purposes. When it comes to streaming solutions, there is still some work to be done for Kubernetes adoption, but applications will support it better eventually. Thus, Kubernetes can provide an application layer that can run in any environment.

11.2 Big Data Systems and Tools

Throughout the book, many different Big Data tools and systems have been discussed. The amount of tooling available in the Big Data ecosystem keeps multiplying, and a Big Data platform needs some of them. It can potentially achieve similar results with a fraction of the systems discussed. When choosing systems and tooling, common sense and a prudent strategy must be used when adopting new technologies. The methods chosen should be solving business problems and getting good results. In this section, previously mentioned technologies and some new ones are discussed.

11.2.1 Storage

The amount of data that needs to be stored for Big Data is substantial. The solutions need to address both ease of access, cost, and resiliency. The storage system or systems should access tools and users to explore data with well-defined access policies. Business pillars should expand the storage by buying hardware or service while keeping costs reasonably low. Storage is the base layer. The storage system chosen should replicate the data and seldom lose it. The system can get offline due to outages but should still recover, ignoring disaster scenarios. Given these requirements, there are two types of storing Big Data: one type is storing it for only offline purposes, and the second type is handling online/offline purposes together. The Big Data platform can use both solutions and have integrations between the two.

11.2.1.1 File-Based Storage

A Big Data platform needs a foundational layer for storing data. File-based storage systems address it suitably. The file-based storage systems are fundamental for data lakes and data intake from many sources. File-based storage systems are durable, cost-efficient, and have granular access controls. There are two prevalent file systems like storage solutions: Hadoop distributed file system (HDFS) and object storages.

HDFS is designed to handle an enormous amount of data. If HDFS is configured to only store data with dense nodes, it can even go higher in the amount of data one node can store. With this setup, HDFS is responsible for data storage, and processing systems should stream data from HDFS to process it. One of the key aspects of realizing such a mechanism is to set up adequate bandwidth for nodes. Bandwidth is essential for recovery and streaming. The applications that stream data from HDFS should not get clogged because of the bandwidth. In case of failure, there is more data to stream back to a replacement node. Thus, bandwidth plays an important role in recovery time. Recovery will always take more time with dense nodes because a single node stores much more data.

The other alternative for file-based storage systems is object stores. Object stores expose a file system like API with directories, client tools, web user interfaces, access control, and integration with many Big Data processing tools. The advantage of object stores is maintenance. Compared to HDFS based solution, object stores do not require much administration other than access control. They also have extra features like object tagging. As discussed in the storage chapter, an organization can leverage both HDFS and object store through connectors. However, the object store should be enough for most organizations if there are no compliance requirements.

11.2.1.2 NoSQL

Horizontally scalable data storage systems are vital for large-scale distributed services. It is ideal for mirroring changes in the production and analytics database immediately in some scenarios. Luckily, some of the NoSQL solutions can easily mirror production data through different replication strategies. Technologies like Spark can connect to the analytics replicas through connector API and execute analytics queries. Cassandra, HBase, and MongoDB are well-known NoSQL solutions that allow analytics processing easily. Moreover, cloud providers also have their in-house solutions for NoSQL databases. In Appendix A, NoSQL solutions such as Cassandra are discussed.

11.2.2 Processing

Throughout the book, Big Data processing was reviewed from different aspects. Big Data processing involves different methodologies concerning the needs of the organization. Organizations might need near-real-time systems to make fast decisions, accurate computation of the data, or both at the same time. As discussed in the previous chapters, all data streams from different sources. The question is whether the system has to answer questions directly to a user request or in a period that is not bound to the context of user interaction.

Depending on the requirements, the Big Data platform might have to offer a solution that uses batch, stream, or a combination of two to serve. For example, pipelines are set up for machine learning, models are trained offline and deployed to production, streaming data is prepared, and prepared data is fed into the trained model to get results. The tricky part is often maintaining a view, understanding, and code sharing among the solution pipeline's different parts. Some techniques for Big Data processing with additional information about combination methods need to be discussed.

11.2.2.1 Batch Processing

Batch processing requires the accumulation of data for a window of time. The window size can vary between minutes to days. For near-real-time systems, the batch window is typically in the order of minutes or hours. For offline operations, the batch window could be in the order of days or more. Batch processing can run on huge data sizes as it can involve days' worth of data. Thus, computation technology should address the needs for recovery and performance.

If the system executes micro-batches, the computing data in memory becomes a natural choice considering performance gains. With periodical checkpoints to recover from failures, the system can be robust enough. It might become important for extensive processing to write intermediate results to persistent storage to avoid massive computation between stages. Technologies like Spark can address both needs, while technologies like Hive are appropriate for huge tasks. Meanwhile, technologies like Presto is a good choice for medium-size data as it cannot recover from failures. Apart from batch processing, stream processing engines can be leveraged for batch processing. Technologies like Flink can handle both stream and batch processing.

Regardless of the batch operation or technology, there are many common problems such as data skew, data type errors, and complexity. The data skew problem can happen due to anomalies, e.g. bots. There are a couple of ways to deal with data skew. One way is to partition data differently or apply techniques like salting. Another way is to preprocess outliers and remove them. For data type errors, a similar approach can be applied. However, if a field is an integer and some portion of data does not conform to integer values, they can be dismissed. Otherwise, the whole batch operation will be rejected. The complexity comes with many ETL pipelines feeding data from many sources. A review of pipelines from time to time to decrease complexity can help, but organizations usually would not have enough time.

11.2.2.2 Stream Processing

Stream processing can address many needs, such as anomalies, fraudulent transactions, and customer experience. Stream processing also depends on window sizes, where the window size is minimal, and its infrastructure generally involves a messaging layer and perhaps a processing engine. Messaging middleware also offers stream processing capabilities. Nevertheless, streaming engines give much more control and flexibility for what one can run.

Messaging middleware is the base for stream processing. Logs, feeds, and events can stream through messaging middleware. Logging solutions like FluentD can publish logs to message middleware. Applications can create feeds for tracking, monitoring, and so on. Besides, applications can create events for various user actions or flows. Messaging middleware solutions such as Kafka or Pulsar can do quick aggregations on these different types of messages. Although simple aggregations cover some surface area, it would not cater to all stream processing needs.

Stream processing engines can combine many streams and generate insights, alarms, and actions based on the aggregations. Solutions such as Heron, Flink, and Spark streaming integrates well with messaging middleware and other data sources. Therefore, they provide many different ways to aggregate data. Stream processing engines can serve a good range of infrastructure needs.

11.2.2.3 Combining Batch and Streaming

Many organizations need both batch and stream processing at the same time. Lambda architecture has been adopted solution for a while to address the combination of the two. Lambda architecture attempts to provide accuracy through batch processing and real-time processing through stream processing. Further discussion about lambda architecture is in Appendix A. Although lambda architecture helps achieve both stream and batch processing, it is hard to maintain due to different code bases between batch and stream processing. Several technologies come to the scene to address both needs.

Stream engines like Flink can handle both stream and batch processing. Batch processing runs on bounded streams, and stream processing runs on unbounded streams. Internally, stream engines optimize performance on bounded streams through algorithms and data structures. On the other hand, they can handle processing events promptly for unbounded streams. APIs for bounded and unbounded streams might slightly differ. Luckily, engines are evolving to provide a singular interface for both bounded and unbounded stream processing.

Another approach for the combination of batch and stream processing comes from abstraction layers such as Apache Beam. As discussed in Appendix A, solutions like Apache Beam offers a unified programming model that can represent and transform data sets of any size regardless of bounded and unbounded streams from any data source. The programming model internally handles both unbounded and bounded data through different components. Moreover, such abstractions can run in multiple backends such as Flink, Spark, and Samza, making them easy to adapt or evolve.

11.2.3 Model Training

Machine-learning algorithms have corresponding implementations on popular machine-learning frameworks such as scikit-learn. The main task is generally setting up good automation for the machine learning model training. With a good workflow setup, machine learning model training can run on the provided environment such as Kubernetes, Spark, and so forth. Once the model gets trained, the workflow might involve steps to push the model to production for consumption. The Big Data platform's main goal is to allow data scientists to train their models and easily productionalize the models.

11.2.4 A Holistic View

For large data systems and tools' solution space, there are a couple of verticals: storage, processing engines, exploration, pipelines, and supporting tools illustrated in Figure 11.1. Depending on the company budget and size, there should be a solution for each vertical to provide a reliable Big Data platform overall. Although there are many alternatives in each of the verticals, it is better to

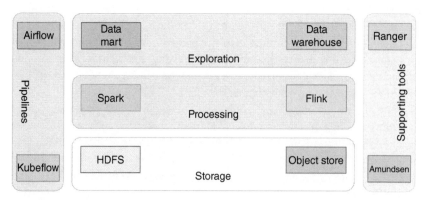

Figure 11.1 Big Data platform verticals.

invest resources consistently. For example, if a company wants to use Spark as a processing engine, they may adjust the rest of the software stack to be compatible with it. However, it does not mean people in the organization should be limited to the tooling that is decided for the Big Data platform.

The storage layer consists of a foundation like HDFS or object store where data can be poured. The processing layer can be tools like Spark and Flink. The processing technologies can run on multiple resource managers such as Kubernetes or YARN. It depends on how the company invests in infrastructure. If the company already has experience in Kubernetes, running Spark on Kubernetes might be easier than building a Hadoop cluster. On exploration, many software choices can be leveraged, such as Presto. Data may be transferred to data warehouses or data marts for exploration and business intelligence. When it comes to pipeline tools, there are good choices like Airflow and Kubeflow. Various supporting tools can be used for security, access control, discovery, and other quality concerns.

While constructing solutions, the harmony of the tools should be observed. The solutions sometimes do not achieve the wanted results. Even then, good integrations with the existing tools and systems should be considered. For instance, most Apache projects integrate well with each other. When building a system or tool, existing projects should be incorporated. Integrations help solutions adopt faster, allows systems to generate value quicker, and expands the product horizon.

11.3 Challenges

Designing a Big Data platform is a long ride amidst many challenges. It entails an understanding of the overall usage, deployment, and maintenance of several Big Data systems. The more components are introduced, the more challenging it becomes. Each new system under a Big Data platform should seamlessly integrate with the rest. The Big Data platform has to evolve constantly. Adding new solutions and depreciating old ones must be done for a better platform experience. In this section, the challenges of designing Big Data platforms are discussed.

11.3.1 Growth

The growth is a nice challenge to have. It means the organization is doing well and the needs are multiplying. Potentially, there is much more budget to spend on the Big Data platform as well. The growth can happen in many dimensions. It is not just the raw data growth but also the number of data sources and the number of systems. Each new data source and Big Data system requires investment in human resources and infrastructure.

Data growth is a natural problem for all growing organizations. The data growth comes from additional services and new customers or users for the organization and requires thorough research on the data's patterns with seasonality effects. More data requires more nodes to multiple systems in the infrastructure. If the platform lives in a cloud environment, it is easier to adjust to spikes. It is harder to adjust immediately for an on-premise solution, as the predictions for the growth should also include a buffer to meet sudden demands. Another matter to consider is the delivery time of new hardware from the day of order since experience tells that order completion can vary a lot.

As the organization grows, there can be more teams, more partners, more acquisitions. It is necessary to integrate with the new data sources to make the best decisions. The new data source integration might involve simple configuration or serious development time when integrations happen through API. Adding more integration points increases operational responsibility and coverage of a Big Data platform. The more resources shared, e.g. YARN, the more contention might happen between different sets of jobs.

Lastly, the organization might need to explore more systems and adopt more tools due to different needs across different organization pillar. The Big Data platform should support new technologies and adopt new solutions. Nevertheless, adopting a Big Data system or solution is not easy. It has to be configured, well-understood, and operationalized. All of these take time and experience.

11.3.2 SLA

Big Data shapes decisions daily, hourly, or even on the scale of minutes. A delay in Big Data processing may cause losses in revenue and inefficiencies. Therefore, delivering data on time is a significant concern for Big Data platforms. Well-timed delivery depends on many factors like downstream data, resource management, runbooks, and human errors. The Big Data platform should address many of these concerns in many directions to limit interruptions.

A large portion of interruptions come from downstream processing delays. Some things can be done to address downstream processing problems. The first step is to identify the owner. If the platform has a discovery tool, then it's relatively easy. The owner can be paged for the delivery problems. Next is to allow partial delivery when it is acceptable. Last is to hold a retrospective meeting for the delivery problem with the owning team.

If the platform depends on a shared execution engine among many teams, then a contention problem caused by one team can result in delays. Luckily, most of the resource managers have ways to separate workloads through queues and namespaces. The Big Data platform should allow easy configuration of workloads. Moreover, preemption can prevent resources from exceeding their limits and provide a fair share of resources depending on the attributed quota.

Setting up service level agreements (SLAs) is a good way to prevent further delays in the processing. Nevertheless, SLAs without actionable items would be rather useless and only wake someone up for no good. For each SLA defined, an associated runbook may be attached to the workflow or discovery tool. The platform should provide the necessary tooling for setting up runbook documentation.

Humans are error-prone, and the platform can offer various solutions to decrease incidents. One of the key issues in data processing is to make changes and apply them to the data. The platform should enable people to make private runs to manipulate a piece of script or workflow easily and test out the results on sampled data. Another good tooling is to have anomaly detection such that the results can be validated. Anomaly detection can detect problems with various simple aggregations.

11.3.3 Versioning

Versioning data is not fun in any part of the software stack. It is not fun in the Big Data world either. Versioning data includes documenting metadata, data validation, schema evolution, and tracking changes on data. Big Data platform should offer solutions to different aspects of data versioning. Nevertheless, there is not a tool that encapsulates all these needs.

The documentation seems to be a burden, but it is an excellent way of written communication. A person looking at a data model may not know the conditions when the model was designed. A piece of information attached to the model can help maintainers or people from other departments to understand the contextual information. In theory, every model should be self-documented. Data definition language (DDL) files for the model should be source-controlled, and models should have descriptions and comments on attributes.

Moreover, additions such as alter commands should also have attribute comments. With this approach, the organization would not have any undocumented model. If the platform has a data discovery tool that can crawl models with respective documentation, people can quickly discover them on demand.

One of the cleanup tasks is dealing with corrupted data. Validating data can partially address corrupted data issues. If there are necessary validation steps in publishing or ingesting data, the cleanup step can be lighter. Validation steps can validate data against different validation rules. Type validation is the most obvious one where the validation step checks whether the input matches the type defined in the schema. The domain might require validation rules that involve multiple fields or custom validation. The critical part is the integration of validation to the rest of the infrastructure. It is hard to implement a central mechanism validating all models as they come from multiple sources. The better validation, the cleaner data the platform would receive.

The evolution of schemas is a technical and nontechnical effort. If the schema updates do not have breaking changes with the older version, updates can happen worry-free. Nevertheless, breaking changes require deprecation and coordination steps. The deprecation steps involve making sure there are no downstream data that depend on the deprecated attribute and use alternative or new fields. The downstream teams should receive updates so that they can apply the necessary changes in time. The platform might offer a discovery tool to provide lineage over fields. Once downstream entities are identified, owners can get upcoming field updates.

Keeping track of changes that happened to data is another challenge. It is particularly important for machine learning tasks. Data scientists often need to version experiments to produce the same results given the same conditions. They want to keep track of data sets and the code that runs the experiments to build the model. Moreover, a data scientist might want to work on the same data for different purposes. Keeping a stable version of data becomes essential for collaboration.

11.3.4 Maintenance

System maintenance is a standard process for applying updates to existing software. System maintenance helps improving performance, addressing security problems, and solving bug fixes. In the Big Data realm, there are many tools and interconnected systems to maintain. Teams build additional tooling or systems for internal use cases. Keeping all these systems and tools in production and updating software is hard. While running an update, many things can go wrong as there is no system, regardless of how small it does not require maintenance and administration. Some of the maintenance tasks can be offloaded to cloud providers or Big Data system solutions. Maintenance has several categories, such as server maintenance, Big Data system maintenance, and tool maintenance.

Server maintenance includes hardware maintenance and OS maintenance. Hardware maintenance requires replacing hardware parts such as disks, CPUs, and cooling fans. The first task for hardware maintenance is the detection of hardware failures through alarms, typically, Nagios. The rest understand the nature of failure, remove the server out of the cluster, order parts if necessary or contact the manufacturer under warranty, and put the server back in the cluster. The second part is the OS maintenance that includes applying security patches, updating system software, and OS upgrades. Ideally, the server can be taken offline out of the server farm, and apply upgrades smoothly, and put it back with others. Nevertheless, the updates might not go as smoothly. Therefore, using updates in a test environment to servers would decrease the chance of having surprises in production.

Big Data system maintenance covers software updates and upgrades to Big Data systems. The most common version of applying updates and upgrades is through the rolling update mechanism. Operational teams generally implement playbooks that execute an update procedure. Teams can test playbooks with virtual environments and avoid potential issues in testing. There are also steps for client tools. With the upgrades on systems, client tools and user machines need updates too.

To ease the amount of time and resources spent in maintenance, cloud providers offer solutions that require much less maintenance through virtualization. Moreover, there is management software like Ambari or Cloudera Manager to help with maintenance, monitoring, and deployment. Some of these solutions also have enterprise plugins and addons to reduce maintenance further by additional functionality. Even with good solutions in the deployment and maintenance, some operations needs to be done, such as integrations and onboarding. The more systems the platform has, the more operational burden it would require.

11.3.5 Deprecation

Big Data is still an emerging technology. There are many tools and systems to replace existing solutions or solve new problems. Although new technologies are getting available, many organizations have outdated solutions. These solutions may still do the job, but they might not receive updates or support anymore. Before adopting new solutions, operational teams need to slowly deprecate existing systems and move off users or systems of the system that is on the deprecation path.

The deprecation process starts with analyzing the existing system and why it makes sense to deprecate in favor of another. The next step is to make a feature comparison between the old system and the new system. Often new systems do not have feature parity with the existing system. The third step is to find dependent clients and systems that use the existing system. With all necessary information compiled, the deprecation process can begin.

The deprecation process can start with notifying existing customers with a clear plan for the deprecation. Many customers might not be happy with the decision since they need to apply migration steps. Nevertheless, it is costly to operate two identical systems without much benefit in a Big Data platform. Maintaining the existing system, even if it is very stable, still requires some effort and resources. Once the operational team is sure of deprecation, they can execute the deprecation. Some people will be late for all announcements, and their system will break, but there is not much to be done. The best way to mitigate is to help them with migration.

11.3.6 Monitoring

Monitoring is a critical piece of Big Data platforms. Monitoring involves keeping track of physical resources, virtualized resources, cloud services, systems running on them, and data hosted on them. It is quite challenging to monitor every system under a Big Data platform and require engagement from the operational team. The more centralized the monitoring solution can be, the easier it gets to have a global understanding of systems and services. Monitoring needs some combination of metrics recording, visualization, alerting, and planning.

Recording metrics from multiple platforms into a central location is quite challenging. Operational teams can leverage tools like Prometheus to record metrics from various components. There are Prometheus exporters for numerous systems in the Big Data area. Cloud providers provide their metric recording solutions. There are also enterprise solutions that help to collect data. Recording metrics can commence using these solutions. One beneficial thing is heartbeat monitoring. It essentially gives simple information about whether a system is up/down. Other helpful metrics are overall latency and load on a system. With a combination of these, three critical problems can be detected easily. However, teams can leverage many other metrics to get a better understanding of the systems in the Big Data platform.

Once the metrics are available in a central repository, the teams can use metrics to set up dashboards, configure alerts, and plan various parts of the Big Data platform. Dashboards can give a quick overview of the systems and components in the Big Data platform. Alerting is necessary to monitor systems healthy and responsive. Lastly, metrics provide a glance at the direction of systems. With this information, management can estimate the cost and order hardware or plan expansion.

11.3.7 Trends

New technologies in the Big Data realm have many advantages since many new technologies are emerging that make the Big Data life cycle more comfortable. On the other hand, organizations cannot spend so much engineering resources on adopting new technologies. Organizations need to be cognizant of the latest technologies while delivering results using existing Big Data systems. Although new technologies are attractive at first, they often come with stability and community problems.

Most organizations do not need bleeding-edge technology for their Big Data platform. Early adopters of such technologies have to deal with many edge cases until the technology gets stable. In many cases, the organization has to contribute back to the technology. While some organizations have a condition to contribute back, many organizations do not have it. Therefore, organizations should closely keep track of impactful changes proven to work for other organizations in most cases.

Moreover, replacing or upgrading existing technology inside the company is not free either. People within the organization got comfortable with one or more systems. Introducing new technology can cause friction if there are no obvious benefits. Even if there are apparent advantages, there are engineering challenges to integrating new technology into the existing Big Data platform. However, it does not mean the platform should not accept superior technologies. The Big Data platform should evolve with better systems and tooling when it is needed.

11.3.8 Security

Security has been discussed from different aspects in a previous chapter. Security for Big Data platforms is about tooling and managing access control on data and planning for breaches and leaks. Data leaks can damage an organization's image but also risk the users or customers of the organization. Credential leaks can allow access to bad actors to take down systems or steal sensitive data. Data breaches can result in an exposure of confidential details through a direct attack on the organization.

Security is not only a problem of the Big Data platform. It is a general problem for the organization. Nevertheless, while building systems and solutions for Big Data platforms, standard security practices should be applied. In the early days of Big Data technologies, security was not the first issue that people tackled. Nowadays, most open-source Big Data systems come with security solutions such as vaults, credential APIs, and integrations to authentication/authorization parties. In designing a Big Data platform, security aspects should be considered.

11.3.9 Testing

Big Data processing involves multiple data sources, data coming in large formats, billions of rows, and the need to make critical decisions. With so many systems are blended, many steps can go wrong. Testing can partly help with error pruning. Several different kinds of testing strategies make systems and produced data consistent, reproducible, and reliable. In a Big Data platform, parts of the machinery can use unit testing, functional testing, integration testing, performance testing, system testing, A/B testing.

Unit testing is well-established for many languages with good framework support. It is still hard to implement unit testing in the context of ETL jobs. The unit test is good for testing the behavior of one component. If the expectations can be set for a given data set, it is possible to unit test the pipeline's behavior. Hand provided data can be placed to the input sources, mocked ingestion part, and tested the expected rows in the outcome table. The execution engine needs to understand schema and table replacement to run unit tests. Ideally, test data should stay in the unit test schema and be ready for other test runs. If tests can be added to prechecks before accepting code, a continuous integration (CI) process with additional automation may be achieved.

The functional tests check that there are no errors in various phases of the big data life cycle. In the data acquisition phase, it validates the completeness of the data. And, various integrity checks can be done to make sure errors are within boundaries. Once the data gets validated, the next thing is to check the data processing and find the overall pipeline outcome. If data conforms to the expected standards that can be tested through anomaly detection, there is a level of confidence the data is reliable. The last part is a data presentation where a final validation check for the outcome table or dashboard is done.

Big Data systems integrate with many components. The components are constantly on the move. There is always a chance that integration with a component gets broken due to various reasons like credential rotation. Integration tests basically should check if the platform is still able to integrate two components. These tests can run on an hourly basis or a handful of times a day to get resolved before a massive backlog of jobs fails because of the integration problems.

Performance testing is important to determine if the system needs additional resources or to find bottlenecks in the platform. The bottlenecks can happen because of data ingestion, data processing, or data consumption out of the platform. Performance testing can happen between many components to get an overview of the platform and test systems individually. System performance testing can give an overall idea of how a given Big Data system performs under heavy load and summarize throughput, etc. End-to-end performance testing can reveal which systems get hosed under stress. Identifying the system or systems can unlock the opportunity to tune them.

A/B testing is the last method that is very useful to see the effectiveness of various machine-learning models. It helps understand the effects of variables in a controlled environment and provides an opportunity to quantify errors correctly. It can help to identify the probability of making an error given the machine learning model. A/B testing requires post-processing to determine bias and statistical analysis to determine the outcome of a test. Data science life cycle tools help with setting up A/B testing. Moreover, some solutions specifically help with the application of A/B testing.

There are many other methodologies and strategies to set up a Big Data platform from various aspects. An organization might not implement all testing strategies but should try to define what is essential and put resources to make sure the testing strategy gets implemented.

11.3.10 Organization

A major challenge in designing, building, and evolving Big Data platform is organization. If the organization does not have a data-driven culture, it is hard to establish key habits and make room for developing a Big Data platform. Thus, it is critical to get an organization on board with Big Data. Organizations may not change easily. The value proposition of Big Data might not be obvious to everyone.

Leadership has to establish the organization's position on Big Data. Even if the organization understands Big Data's value, it may lack decision-making with Big Data, a common strategy, and a unified approach.

Embedding Big Data into daily decision-making activities require changing of habits. Some organizations inherently are better at it because they have been data-driven organizations from the establishment. In many cases, incorporating Big Data into daily routines might need executive sponsorship. Leadership can promote a data-driven approach to key decisions. When people get positive results with data, an organization can celebrate achievements so that data becomes a norm for actions.

In many cases, organizations might want to use Big Data for many decisions. They can start building pipelines and make sense of it in some way. Different business units can make disparate attempts to get the best of the data. Without a common strategy, organizations will spend resources on similar tasks often. A couple of systems wired together in a different part of the organization might not bring a Big Data platform into life. Therefore, organizations need a common strategy and build key infrastructure for Big Data platforms.

A common strategy brings a unified approach and centralizes some of the infrastructure pieces into the platform. The infrastructure team can design, deploy, and maintain core pieces of the platform, while business units can leverage the platform and support the infrastructure team. A unified approach can help increasing efficiency across different business units. Different teams would encounter the same problems and can do knowledge transfer. With a unified approach, teams can better master the systems that the organization decides to maintain.

11.3.11 Talent

Big Data requires a diverse set of skillsets like communication skills, analytical skills, software engineering, operational experience, statistics, and business acumen. In addition to these skillsets, designing a Big Data platform requires understanding different Big Data systems, how they integrate, how they work together, their pain points, and vision for evolution. Although engineers can acquire some of these skills through self-studies, they have to learn a big portion of it on the job. The experience is decisive since real-life production experience allows dealing with edge cases, failures, and sometimes catastrophes. Finding talent who has some of these skills and retaining them is a big task.

A Big Data platform team needs a different set of skill sets. One person who has all the required skillsets is not a common scenario. It is rather important to find talent who complements each other. One team member can be good at security, and another at automation. Teams can design, deploy, and maintain Big Data systems with overall wisdom. Effectively, management should target skill sets that do not exist or are not sufficient to handle the team or teams' work. Management has often sacrificed one or more skillsets for talent. It is hard to find a talent who perfectly fits the environment. Hence, organizations should look for a good match with the learning path.

Retaining the existing talent has very complex dynamics and out of the scope of this book. Nevertheless, having an environment to learn, try, and fail helps engineers excel at what they do. Good engineers take pride in what they do, and giving them enough freedom with supporting activities is another way to help. Teams that work well are a huge asset. Spending time and budget on working teams is never a loss but a considerable gain in the long run.

11.3.12 Budget

Budget is a defining factor for many of the systems or talent for Big Data. Although it is nice to have the flexibility to use any technology or hire talent, organizations are restricted by budget in reality. The budget limits training for existing staff, recruitment, external consulting, and investment for additional services or resources. An organization has to keep ongoing Big Data systems as well as introducing new ones.

The challenge for a Big Data platform is to prioritize tasks. Every system or team might seem to need many new resources. However, many savings can happen under scrutiny. There are two major ways to cut down the cost: automation and optimization. Many procedures need to run for part of infrastructure maintenance, provisioning, and so forth. The more procedures a team can automate, the less work needs to be done. However, many systems can be tweaked to run with less memory or CPU with proper configuration. With a combination of the two, the running cost of a Big Data platform can be lowered.

12

An Ideal Platform

After reading this chapter, you should be able to:

- Understand the use of event sourcing for Big Data
- Understand Kappa architecture
- Learn about shift toward data mesh
- Learn about the use of data reservoirs
- Describe data catalog for the platform
- Understand the need for self-service platform
- Learn about Big Data abstraction
- Learn about design trade-offs
- Learn about data ethics

Throughout the book, many great technologies and approaches have been reviewed. The summary of what we have talked about before is found in the previous chapter. In this study, the challenge has been coming up with a simple platform while keeping it efficient in terms of development efforts and running costs. Designing such a platform is subject to many constraints such as talent, growth, size, and budget. There is no one-size-fits-all solution for a Big Data platform. A solution that works for an organization might not apply for another one. Therefore, some of the effective patterns for designing a Big Data platform will be studied here. By synthesizing technologies with patterns, a Big Data platform that delivers tremendous value to the organization can be obtained.

It is believed that designing a world-class Big Data platform starts with early integration to the rest of the services. If the ingestion is lagging, then we have more troubles in working with the data. The load of work increases from cleansing, processing, maintaining, deploying, and so forth. The earlier the Big Data platform is integrated to the rest of the infrastructure, the easier it gets to handle the journey of Big Data. Thus, the suggested patterns treat Big Data platforms as the first-class

Designing Big Data Platforms: How to Use, Deploy, and Maintain Big Data Systems,
First Edition. Yusuf Aytas.
© 2021 John Wiley & Sons, Inc. Published 2021 by John Wiley & Sons, Inc.

citizen of everything a company does. When the organization has a new feature or product, it needs to consider how data flows through various phases of processing.

The approach suggested is treating data as a product and an asset. To deliver results for a product, one needs to think about the market, stakeholders, roadmap, business, and so on. When adding a new product or feature, people often discuss some details before the application, such as data, its model, storage, retention, customers, implementation strategy, SLO, etc. The proposed patterns strongly correlate with data as a product and assist in establishing a successful product.

12.1   Event Sourcing

Event sourcing is a pattern that captures every change to the application state. Applications that use event sourcing log every change event. The application state can be reconstructed from event logs at any point in time. Event sourcing provides a stream of messages that can be leveraged in the Big Data processing. The organization does not necessarily need to use event sourcing for everything, but passing down events, models, or messages to the Big Data platform for further processing gives out many opportunities. Let us quickly go over event sourcing and describe it.

12.1.1   How It Works

The idea behind event sourcing is capturing state changes. Instead of directly saving the application state to data storage, the changes get published as events. Let us consider a simple example where an e-commerce company processes an order. Traditionally, the ordering process has several components. In a world of microservices, we can think of order service, shipping service, payment service, and so on. In the following diagram, several events happening one after another can be seen (Figure 12.1).

If we were to save state changes directly to a database, we would have a shopping cart table for a customer. The following queries would occur as a result of the operations above:

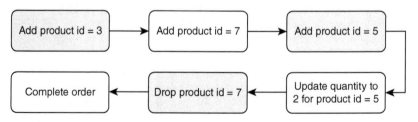

Figure 12.1   Shopping cart events.

```
INSERT INTO shopping_cart (cart_id, product_id) VALUES(29, 3);
INSERT INTO shopping_cart (cart_id, product_id) VALUES(29, 7);
INSERT INTO shopping_cart (cart_id, product_id) VALUES(29, 5);
UPDATE shopping_cart SET quantity=2 WHERE cart_id=29 AND product_id=5;
DELETE FROM shopping_cart WHERE cart_id=29 AND product_id=7;
UPDATE shopping_cart SET order_completed=true WHERE cart_id=29;
```

If we were to have the same actions logged as events, we would arrive at the same last state. Instead of having SQL queries above, events that represent state changes would rather be obtained, which would look like as follows:

```
[
  {"eventType": "AddedToShoppingCart", "cartId": 29, "productId": 3},
  {"eventType": "AddedToShoppingCart", "cartId": 29, "productId": 7},
  {"eventType": "AddedToShoppingCart", "cartId": 29, "productId": 5},
  {"eventType": "UpdatedProductQuantity", "cartId": 29, "productId": 5,
   "quantity": 2},
  {"eventType": "DroppedProductFromShoppingCart", "cartId": 29,
   "productId": 7},
  {"eventType": "OrderedShoppingCart", "cartId": 29}
]
```

Event sourcing suggests keeping each change as an immutable event that leads to the final state instead of a mutating state. The problem with the mutable state is the loss of history. We can always recompute the state from immutable events. Representing the state through immutable events gives more detailed information. Keeping immutable records over mutable changes is the core of event sourcing (Fowler, 2005). There are some implications for event sourcing:

- At any time, the application state can be rebuilt by applying events historically from the event log.
- The application state can be determined at a snapshot in time. A snapshot in time data might be particularly useful for machine learning.
- Writing and reading of data differ as reading needs an aggregated view of events.

12.1.2 Messaging Middleware

Once events were constructed for every change, they need to be stored somewhere. An event can correspond to a log entry or a message. Therefore, we can naturally use messaging middleware to store events. By introducing messaging middleware, we can construct a state through aggregation but also distribute events to any interested party. In the shopping cart example, one interested party might be a shopping cart service. Another interested party might be a product recommendation service.

With event sourcing, everything can be derived from raw events. There can be a process like shopping cart service that derives aggregates from the raw events and updates the caches when new events come in. In the shopping cart example, shopping cart service can simply create a caching layer using a database either in memory or disk to serve a cached view of the event log. On the flip side, the recommendation service can feed raw events to a machine learning model to update recommended products for the customer near-real time. The good thing about raw events and the cached state is denormalization (Kleppmann, 2016). The shopping cart service can potentially have other events and give a consolidated view of the shopping cart for the customer.

With messaging middleware, all state changes can practically become a stream of events. Basically, event sourcing facilitates stream processing. With a stream processing engine, we can move much of our Big Data processing to stream processing with potentially several time windows for varying business needs.

12.1.3   Why Use Event Sourcing

Imagine we take a dump of the shopping cart table. We would not have an entire picture for the customer while he/she is shopping. On the flip side, if we would go with the event sourcing approach, we would have an entire history of everything the customer did and readily available for further processing. With messaging middleware, the events can be partially stored in the messaging infrastructure for a time span. Later, tiered storage or offload data can be potentially used to Hadoop Distributed File System (HDFS) or object store for longer retention.

Adopting event sourcing also has new good side effects. Since a messaging middleware is introduced, we can potentially invest in different directions in the same way. Take visitor tracking. Many websites want to know visitor's history through the website such as clicks, impressions, views, and so on. All these user actions can be recorded as events such as PageViewEvent or ClickEvent. Instead of getting such information through log parsing, these tracking events can be directly published through the means of event sourcing. The advantage is precise control and schema. The log parsing can be hairy and might be harder to debug. What is more is that we want to treat data as a product. It is easier to integrate a tracking model than parsing unstructured logs.

12.2   Kappa Architecture

With event sourcing, we have converted every change into an immutable stream of events. We can use a stream processing engine to process these streams immediately after they become available. Kappa architecture depends on principles that are easy to accomplish with event sourcing. It has the following principles (Kreps, 2014):

- Batch processing converges into stream processing. It is simply special cases of stream processing where the data is bounded.
- Streams consume immutable events from multiple sources. We persist raw events and compute the state of the application from events. Views of raw events can be computed in real time or offline.
- The implementation of Kappa architecture utilizes a stream processing engine for both streaming and batch operations. Having one system for both stream and batch processing significantly decreases code, maintenance, and operations.
- Since we store raw events, we can replay whole logic again in case of changes to the business logic. Thus, the replay mechanism allows evolving previous computations by the historical data from one or more streams (Figure 12.2).

12.2.1 How It Works

To bring Kappa architecture to life, a messaging middleware is needed that can store the data for a certain period of time. Moreover, the messaging system should provide a mechanism to retrieve data starting from an offset to enable the replay mechanism. Both Pulsar and Kafka are good choices for messaging middleware. Pulsar supports tiered storage. It enables storing more data by offloading some of the data to HDFS or an object store. Once we establish messaging middleware for the organization, we can find ways to make everything a stream of events.

Event sourcing enables a stream of events naturally. Thus, it would be the preferred approach to store data. Moreover, we can also treat other data such as logs as stream and potentially use a solution like Fluentd to stream data directly to messaging middleware. Depending on the circumstances, logs can be parsed either before or after publishing them to messaging middleware. Once everything streams into messaging middleware, the next thing is introducing a stream processing engine.

Stream processing engines can work with different windows. The same code can be run with different window configurations if necessary. Depending on how the result is used, computed results might be published to messaging middleware for further processing or storing in the data store. If processing code changes, there is a need to recompute results. Recomputation runs on the same infrastructure with the same input. For stream processing, we can use Apache Flink or Apache Spark, which both work well with streams and provide enough capacity to develop complex pipelines. The last step is storing computed data.

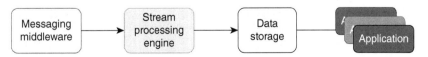

Figure 12.2 Kappa architecture.

We can store computed results in different tables depending on the configuration. A typical streaming table can store results in partitions. Once a window completes, it should save the respective partition to the table. For many applications, direct stream processing output might be good enough. Perhaps, there is no need of better precision. In that case, the streaming table can be a general-purpose table. When precision is needed, then results can be saved to a different table for applications that need more accurate data.

12.2.2 Limitations

Kappa architecture is a great way to cater to a large portion of Big Data processing. Many applications in an organization can depend on the output of Kappa. Though, Kappa is not the answer to all Big Data processing. Organizations may still need additional pipelines to process data. It is assumed that Kappa is the frontier of the overall processing and makes it easy to process Big Data further. Although the claim is to store data in messaging middleware for long periods of time, it can get quite expensive to do so. Multiple data sources with different pace can result in problems because of accumulation. Offline data processing that does not require streaming at all would be more complex to implement in Kappa than traditional batch ETL pipelines.

Messaging middleware solutions like Kafka or Pulsar can store petabytes of data. Nevertheless, messaging middleware is not a data reservoir. Storing data permanently or very long periods of time can be expensive. It is also much more complex to store vast amounts of data compared with an HDFS or object store. Partitions, partition size, node disk size, etc. need to be considered. On the flip side, adding nodes to Hadoop or simply using more space in the object store can be cost-efficient and reliable.

Streams can come at a different pace. To compute the window, one stream needs to accumulate more data to process them together. Accumulation can cause memory problems. Moreover, the more streams combined, the harder it becomes to understand. Results can be published to middleware, but it helps to a certain point.

Some of the processing does not really need any stream processing environment. We can still use a stream processing engine; however, it does not fit Kappa architecture. The Big Data platform might receive data with significant delays in the order of days. As an alternative, a dump data can be received from one or more resources. There are many ways where data might not be naturally a stream. In those cases, there is not much sense to treat them as a stream.

12.3 Data Mesh

In the last decades, applications have transitioned from monoliths into distributed and domain-driven services. Many organizations in different businesses adopted or are still adopting the modern approach. Services provide isolation, productivity, flexibility, and scalability for organizations. The same kind of movement might be highly beneficial where data is naturally distributed and domain-driven. In a sense, organizations have been building their data monoliths. Many have embraced a centralized team that tries to implement various data processing tasks. Big Data platforms may suffer from the same type of problems once applications suffered.

As the number of data sources and customers for data proliferates, it becomes increasingly hard to have everything under control. A centralized infrastructure can quickly become a bottleneck for the organization. Coupling irrelevant data with different customers might hinder release time for new data products. Moreover, working on the same data pipelines for different purposes can slow down the release time. It is also against the notion of architectural quanta, where components are highly cohesive and independently deployable (Kua et al., 2017). A monolithic approach ends up in highly specialized but siloed teams. People who work with data day to day have to own broad responsibilities to get things done. Communication problems occur between different teams due to the loss of information between siloed teams (Dehghani, 2019).

A paradigm shift toward decentralization can help with the shortcomings of the monolithic Big Data methods. Moving to a domain-driven, self-serving, and data as product strategy might overcome deficiencies of monoliths. A decentralized strategy might improve delivery time, organization memory, and autonomy.

12.3.1 Domain-Driven Design

The domain-driven design (DDD) changed the way engineers approach designing software systems. DDD concentrates on deeply connecting the implementation to the business concepts (Evans, 2004). Although DDD practices are widely adopted in various parts of software development, they have not been the focus of Big Data platforms. Big Data platforms are a natural habitat for DDD practices as they turn data into actionable business insights. The core of getting results for Big Data is capturing the domain correctly. DDD comes with many great ideas for decomposing complex business applications into modularized services. Applying the same approach to Big Data platforms can help with reducing the complexity of monolithic issues.

One of the core patterns in DDD is bounded context, which is a logical boundary between subdomains in a business domain. For an e-commerce application, bounded contexts can be shipping, payment, and so forth. Bounded context provides modularization of applications. Modern services use a bounded context to define the boundaries of service. In the same way, ownership and data sets that strongly align within a logical boundary can be defined. Each subdomain, team, service, or collection of services should own their respective data set and serve them. Interested parties like cross-domain applications can then subscribe to the data set supplied by the owner (Figure 12.3).

Instead of coupling all data domains in one monolith, bounded contexts can be used to define ownership of data and its relevant pipelines. Capturing data at its origin can deliver subdomain facts at its source. Having local pipelines for subdomains can relieve the pain of tangled pipelines and provide autonomy for the service and data set. The team that owns particular services should also own the relevant data set. In this setup, if we use an event sourcing approach with Kappa architecture, we can then stream domain events. Subdomain pipelines, data flows for services that the team owns, can retrieve domain events and process them to create source domain data sets. Each subdomain data set should provide service level objectives (SLOs) such as delivery time, error count, and so forth. The organization can then use source domain data sets for business intelligence and data-driven decisions. Cross-functional data products can use more than one data source to answer additional questions.

Since each subdomain delivers data to its respective data set, products that require more than one domain need to either pull domain events where it is applicable or consume the provided data set. The trick here is to establish a Big Data platform that can easily move data around without much effort. This approach provides clear ownership of upstream data sets that would increase the productivity of downstream teams.

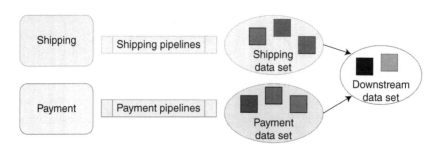

Figure 12.3   Subdomain data sets.

12.3.2 Self-Serving Infrastructure

When data gets distributed over multiple systems and subdomains, there is a risk of wasting resources. The key to minimizing duplication of efforts or resources is enabling self-serving infrastructure with minimal support for subdomain processing and storing. The ease of use for various infrastructure components is an important factor for self-serving. Friction during development, deployment, and maintenance can cause hacks and misuse of infrastructure components. A Big Data platform that serves many pillars from different parts of the business has to provide agnostic solutions to support common data tasks.

Organizations adopting distributed data solutions may have the same theme for various tasks. For instance, streaming domain events, processing them into data sets, and serving them for consumption are common activities. The infrastructure should provide enough capability for individual subdomain teams to perform common activities. On the flip side, the flexibility for new systems or components is highly decisive as different subdomains might require different tooling. For instance, teams might want to save the resulting data set to ElasticSearch or Neo4J with respect to their needs.

12.3.3 Data as Product Approach

In the world of services, each service solves a particular business problem and exposes itself to the outside world through application programming interfaces (APIs). To complement consumers of these services, services are often discoverable and provide documentation on the usage of the service. Decentralization brings challenges like discoverability and proper documentation to the data world, too. The same rules apply to the data and data pipelines when they are decentralized. Therefore, the organization needs to change the way they perceive data. Data should be a product and treated as a product that evolves with the organization. Downstream consumers of the data like data science and business intelligence should be treated as customers.

When data becomes a product, then data product needs to meet SLOs. Delivery time, correctness, volume, and more should be part of the data product commitments. In data as a product approach, the model that is exposed to customers becomes the interface. Data product owners need to be vocal about the product evolution. Any interface changes or deprecation should be a notification for customers. Moreover, the models should have proper documentation and should be easily discoverable for new or existing customers. Data products should also have a standard definition and should be served in a standard format to make it easy to consume. Lastly, onboarding documentation for the data product would add value and might bring additional customers.

12.4 Data Reservoirs

Organizations have adopted data lakes as their primary repository for all data. Nevertheless, many organizations ended up as data swamps instead of data lakes due to lack of organization, policy, and standards. Despite the immense volume of data in data lakes, a little amount of data actually makes it to production. Instead of adding value, much of the data streams to data lake pollutes it. However, dumping everything into underlying data storage was acceptable because the storage itself is not expensive in theory. In practice, it gets quickly expensive. Moreover, organizations have been fighting against pollution by tasking operational teams to find unused or unidentified data. When data is streamed into the underlying storage with well-defined, well-controlled, and well-managed pipelines by teams, pollution can be prevented.

When the organization is stopped from polluting the storage layer, a well-maintained data reservoir can be formed. Preserving a data reservoir takes organizational steps to prevent pollution. Organizations can benefit from a healthy storage layer where data is organized and has policies and standards. Stopping pollution can be a good step forward; however, having completely irrelevant data might be another obstacle. Seldom one huge data storage itself can become a hindrance. Perhaps, one data reservoir does not have to cover all organizational data. Depending on the needs, the organization may also use multiple data reservoirs (e.g. one for each business vertical) to address different business requirements.

12.4.1 Data Organization

Organizing Big Data can help with both the operational and physical costs of maintaining a data reservoir. Organizing Big Data involves reducing data footprint, controlled data collection, and setting up a business goal for each data set. The first step is to bring down the vast amount of data stored. A big portion of Big Data has no use for data analytics or data science. Cutting down the irrelevant data can help with operations and reduce physical cost. If the overall data has cut down data footprint, then it can store other valuable data. At the same time, less data means faster response time for queries scanning through the data set.

A second step is to control what gets collected. Since data lake has been introduced, organizations started to dump any data into data storage. The consequence of dumping data without purpose is inefficiency with additional cost. A better way is to allow teams to collect data but ask for the business objective. Besides, it is possible that the organization does not need all the data. There might be opportunities to prune data to make it both more understandable but also storage-friendly. And the last step is to align data with the business service or pillar. If the organization follows the data mesh paradigm, alignment happens naturally. Each business service or team can manage their own data. Alignment also helps with control since the source of the data can be traced easily.

12.4.2   Data Standards

Establishing standards for the data reservoir is essential for productivity, efficiency, and evolution. Raw data can come in many different formats without proper documentation or schema. Storing raw data as it comes has unfortunate consequences, potentially polluting ones. Storing raw data is not optimized for space or processing. The raw data may lack description and can easily become a burden rather than a win. Evolving data might have unexpected outcomes as the data may lack integrity.

A fair amount of data come in simple formats like comma separated values (CSV). Often, these files do not have any description or documentation for what it carries. Pouring such data directly to the reservoir will quickly pollute it. The data needs to be cleaned, documented, and put in a space-efficient format like Parquet before it gets to the reservoir. This might seem counterintuitive. Nevertheless, standards should be enforced to avoid pollution. If data stays in the format it comes, it will consume more space. Each time it needs to be processed, additional resources will be used. Moreover, if the data does not get clarification about what it entails, then nobody in the organization will have any idea about the data after some time.

12.4.3   Data Policies

Data arrives at the data reservoir from many systems. A part of the data might be more sensitive than others or can require different access controls. Some data might be retained for a period of time. The data reservoir needs access, retention, and business policies. Lack of policies might risk the organization's brand and open the data reservoir for potential misuse and degradation.

Access policies are necessary for any application with sensitive data. Data streamed into the reservoir should get classification, ownership, and other access features. There can be several ramifications of data without proper access policy. The team that owns the data set should manage access policies. Moreover, each part of the data should have a retention policy. Accumulation of data can be problematic in terms of storage and performance. There can also be legal implications of retaining data more than a predefined period. Lastly, data should have a business policy that defines how, for whom, and what to share.

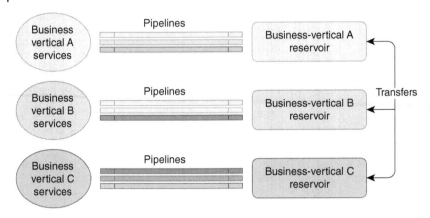

Figure 12.4 Multiple data reservoirs.

12.4.4 Multiple Data Reservoirs

More than one data reservoir approach has not been much discussed; however, large organizations can benefit from multiple data reservoirs. An organization that has several verticals can have separate concerns for each vertical. Each vertical might have its own unique data sets. It might not make much sense to have them in the same repository. The approach aligns with data mesh where each subdomain might have its own pipeline and end up in its own data reservoir. The concern might be duplication of efforts or transferring data between data reservoirs. Both concerns can be addressed through self-service tooling (Figure 12.4).

There are a couple of advantages to multiple data reservoirs. Ownership of the data reservoir would make teams critically think about the cost. Each department can easily find out their overall cost and try to be cautious when spilling data to the reservoir. On the flip side, some verticals might not even need a data reservoir. They might pump their data into a data warehouse or data mart. Besides, the data sets in the reservoir can have a much tighter business relationship as it directly belongs to the services and team that use it. Lastly, some verticals might want to use a specific cloud provider with their object store. Others might want to use an on-premise data reservoir or another cloud provider. Providing flexibility in the data reservoir can improve efficiency and productivity.

12.5 Data Catalog

When many components in a Big Data platform are distributed, it gets even harder to reason about data, lineage, ownership, business purpose, and so on. Many disparate systems require the centralization of knowledge to connect the

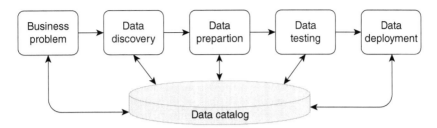

Figure 12.5 Data catalog feedback loop.

dots between them. Thus, data catalog becomes a vital part of the Big Data platform. The data catalog collects, presents, and visualizes data and its journey through systems. Each system in a Big Data platform should integrate with the data catalog to serve information. Every new system that comes to life should also register itself with the data catalog. The catalog would then contain all relevant information for all customers and empower people or systems to discover data efficiently.

The lack of centralization of organizational knowledge might end up in siloed information among different systems. The organizational memory might depend on tribal knowledge or scattered to multiple wiki-like software. On the other hand, a data catalog compiles information from many sources. It helps people and systems through the journey from finding data to solving a business problem. Each step in the chain of activities both contribute back to the catalog and receive the information they need. Establishing the feedback loop with the data catalog requires ease of integration and registering each and every component with the catalog.

12.5.1 Data Feedback Loop

Developing a data product has many phases. Each phase needs an understanding of the data. A data catalog can create a feedback loop to help people and systems. People can see all the data they need with the information and background story. The contextual information with mapping and linking can lead to significant efficiency. A typical data project starts with a business purpose and then a discovery of what is at hand and completes with preparation, testing, and deployment. In each step in the development, the data catalog can serve and receive information (Figure 12.5).

The feedback loop also depends on how much information people supply to the catalog. It should be an organizational effort to keep the information up to date and never leave a bit of data without decent documentation. If an organization practices good citizenship for data documentation, a data catalog can provide confidence, efficiency, and speed to both people and systems.

12.5.2 Data Synthesis

Integration to almost all systems is pivotal for a successful data catalog. There are two ways that the data catalog can integrate with existing systems. The first way is through crawling an existing data source. The second one is through an API where external systems would publish data changes to the data catalog. Collecting all metadata information in one repository is highly beneficial. If the metadata contains additional information such as partition columns and data types, then systems can use metadata to do transformation or some other practical application. With API integration, applications can publish metadata updates at its source. The quest does not end by just accumulating the data. A data catalog can synthesize data in many different ways such that it becomes easier to consume and reason about.

Data synthesizing includes grouping, clustering, and indexing metadata in ways to show a consolidated view of overall data scattered around many systems. The more effort organization puts in synthesizing, the easier it gets to reach information. The data synthesizing gets more important as the organization grows as the number of data sources increases. The same type of data can appear in different systems without any way to tell where they come from. Data synthesis techniques can deliver a broad range of visualization and relationship mapping. I believe it is a critical area to invest resources to get the best out of data.

12.6   Self-service Platform

Establishing a self-service Big Data platform that is agnostic to business logic can improve organizational effectiveness in many directions. The lead time to deliver a data product can decrease significantly. People who work with Big Data daily can forget about the complexity of underlying systems and focus on business problems. On the flip side, Big Data platform engineers can focus on platform enhancement, maintenance, and support. Many Big Data solutions claim to provide a self-service system for Big Data needs. In reality, a self-service system requires a lot of automation and configuration. Most systems still require a lot of engineering resources to make it close to self-service. Besides, building a self-serving platform gets trickier when multiple systems need to blend.

Organizations can outsource a large part of automation to cloud providers. Platform engineers can then spend less time on operational activities. Yet, there are

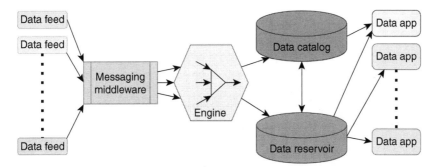

Figure 12.6 Self-service Big Data platform overview.

still many aspects that a Big Data platform has to automate. The self-service experience starts from the ingestion or publishing of data until the end of the data product pipeline. In each phase, people should plug the Big Data platform into the business context seamlessly. In an ideal world, the Big Data platform offers self-service capabilities to data publishing, data discovery, data processing, data monitoring, and more. The platform engineers work on these capabilities and make each element self-serving. Let us go over some of the self-serving capabilities.

12.6.1 Data Publishing

Data publishing activities start from the source. When a business line wants to ingest data models into the Big Data platform, they should be able to create a data model in a polyglot way. Each new data model becomes a data feed that addresses a specific business objective, such as tracking click activities. The Big Data platform should then receive the data, register it with the data catalog, publish it to the relevant topic. The downstream consumers of the data should be able to discover data and start consuming it (Figure 12.6).

The data feed might be an integral part of a self-service Big Data platform. The Big Data platform can provide an API for registering and ingesting new feeds. Underneath registry operations, the Big Data platform can save schema, handle versioning, publish metadata to data catalog, and potentially prepare messaging middleware, such as topics. With this setup, a business line can add a new feed on their own without minimal support from the platform team. There are though cases such as breaking changes to the schema that might require some coordination. Nevertheless, setting up a new data product can become remarkably straightforward. Note that data feeds can come from other sources, such as logs or domain events. In this approach, everything becomes a data feed that gets published to messaging middleware.

12.6.2 Data Processing

Data processing can turn into a self-service experience for mixing many data sources into different aggregations once data becomes available. In a large-scale Big Data platform, there can be a variety of tools, systems, and connections between them. The Big Data platform can provide tooling to set up pipelines and integration with many data sources and productionalize data products. Platform engineers can work on wrappers, abstractions, and automation for different systems to simplify the experience of platform customers.

The Big Data platform can provide wrappers to make it easy to transfer data between data sources, aggregate various data, and handle automation. On the background, the wrappers might register metadata, check permissions, provide logging, version models, collect metrics, and so forth. The platform user then can only focus on the business problem. With such a tailored experience, the platform can keep track of mundane tasks while customers can focus on delivering value. Abstracting away from such details can also bring standardization over the systems. Platform customers can use standard places to look for logs, metrics, and visualization of phases of processing. The overall process of developing, debugging, and productionalizing can get easier. The platform engineers can also provide standard setups for common tasks to further improve efficiency.

12.6.3 Data Monitoring

Data monitoring involves detecting anomalies to any disruptions to underlying data pipelines or data operations. The Big Data platform can offer various self-service tooling to make it easy to detect, act, and solve data problems. The platform can offer the necessary tooling to detect issues, such as latency, corruption, or loss of data. Moreover, the platform can do passive monitoring like versioning, backward compatibility, and so on. In the event of issues, the platform can notify interested parties. Platform engineers can implement such capabilities as part of the platform support and improvement.

12.7 Abstraction

Most of the Big Data systems depend on distributed computing. Distributed computing can get quite complicated in many cases. It can make it harder for starters to enter the realm of Big Data or struggle after working with Big Data. Interconnected components might reveal so many details without limited benefit. A person who only needs to write an exploratory query can easily be overwhelmed. The Big Data platform can provide abstractions over systems, integrations, and workflows to address these issues. Abstractions can significantly reduce the learning curve and provide a smoother experience to platform customers.

Abstractions can emerge in different ways. A Big Data platform can implement user interface abstractions for everyday tasks or provide libraries or API that hide the complexity of underlying systems. Right abstractions can significantly decrease ramping up time for users. Nevertheless, abstractions should not steal away the ability to tune or customize underlying tasks. Advanced users should still be able to control their jobs and environments in many aspects. Abstractions should provide the magic for novice users but have an unboxing option.

12.7.1 Abstractions via User Interface

User interfaces are great ways to make the user experience trivial while keeping the essence of the operations. A Big Data platform might have many interconnected components. Between those components, it can become hard to navigate or often visualize. Abstractions over these components can help to comprehend the state of the data and achieve ordinary operations without much complexity.

Visualizing the overall state of the processing can get tricky. Each subsystem might have their own user interfaces or APIs. Creating a bridge between these subsystems and centralizing the view state can deliver a consolidated understanding. Thus, one of the areas where platform engineers can focus on is state reporting. The better the state reporting gets, the easier to track down problems. Ideally, the summary view should link to the subsystems to give a more detailed understanding. Workflow tools generally have these concepts; however, not everything lives in workflows. Organizations might need some extra effort to visualize the overall pipelines.

Common tasks like data transfer or aggregation might require setting up pipelines, installing Big Data tools, developing data manipulation tasks, understanding of complex Big Data systems. Nevertheless, it can get quite hard for non-tech-savvy people to get something done. Hence, the platform engineers can provide tooling that does operations through the user interface. With the correct setup, anyone in the organization might be able to consume Big Data in a self-service manner. The Big Data platform can support as follows:

- Scheduling data transfer tasks through the user interface.
- Running a query and saving the result in a shareable fashion.
- Adding scheduled aggregations that send report.
- Creating alerts that send a notification to subscribers.

These are some examples of abstractions through the user interface. Users might not need to know how these jobs get scheduled or run. The platform tooling can figure it out for the user. The tooling can offer a generic interface for each one of these jobs. The tooling then can painlessly create actual jobs on target systems without user involvement.

12.7.2 Abstractions via Wrappers

Wrapping libraries, APIs, environments, and systems can provide a simplified model for Big Data processing where the platform users can concentrate on the business logic. There are ongoing efforts like Apache Beam from the Big Data community to add abstractions to different layers of Big Data processing. The Big Data platform can use these abstractions to provide a better experience. When there are common pain points that people deal with daily, platform engineers can wrap these libraries, APIs, and systems to provide a smoother experience.

Wrapping libraries or APIs can decrease the mental effort to get something done when users only use a portion of the library or APIs. Wrapping libraries and APIs also give a chance to easily switch to a new implementation without platform users noticing underlying changes. Moreover, the platform wrappers can provide additional methods that are widely used in the organization. These additions can become a community effort within the organization. People can add new functionality based on their needs when they are common enough.

Wrapping systems and environments abstract away how the particular job, code, or pipeline runs. The platform user can implement business logic without details of how it runs. The platform can then accept the implementation, convert it to a deployable artifact, and run on a target system that the platform supports. Abstracting systems and environments can allow replacing an existing engine with another one. Moreover, the platform may even choose a runtime depending on the job dynamically. For instance, if the current cluster is running at full capacity, perhaps, the platform might spin up a cluster on the cloud or Kubernetes. In consequence, removing the link between systems and environments can ease user experience and provide flexibility to the Big Data platform.

12.8 Data Guild

Coming up with a Big Data strategy should be an organizational effort. A data guild is an open forum or a recurring meeting for Big Data for the organization to define and evolve its Big Data strategy. The members of the data guild can discuss various aspects of Big Data management within the organization and guide teams in their Big Data solutions. When most of the Big Data efforts are distributed over multiple teams, solutions can diverge from the organizational strategy. The data guild can help organizations to prevent disparities between different business units and evade associated inefficiencies.

When different business units control their data products, they might have different approaches and apply diverging patterns. Although individual teams should be free to choose their own Big Data strategy, having excessive differences can cause inconsistency and flaws. Fortunately, the data guild can set up guidelines for Big Data development to avoid unpleasant ramifications. The Big Data guidelines can roughly describe tenets for Big Data development for the organization. Individual teams can then base their principles on the organizational guidelines and extend as they see fit. Teams can also contribute back to the guidelines through data guild events.

Data guilds can review various designs from individual teams to avoid wasting resources and making the best out of organizational talent. Teams might not be aware of solutions other teams implemented for a similar problem. Data guild can help individual teams to identify existing solutions or approaches. Moreover, data guilds can assist in designing well-rounded Big Data solutions with inputs from different individuals. By bouncing ideas internally, teams can also understand how Big Data solutions might affect other teams. Last but not least, the data guild can organize an agenda for main Big Data events such as adoptions, migrations, deprecations, etc. Data guilds can deliver an organizational roadmap for Big Data and minimize the disruptions for the Big Data consumers.

In consequence, the data guild is a central communication channel between individual teams. These are highly beneficial to bounce ideas and define a Big Data strategy. Data guilds can build consistency between teams and support teams in designing next-generation Big Data products. Data guilds can support streamlining data products and help the organization come up with world-class Big Data solutions.

12.9 Trade-offs

Designing Big Data platforms requires many decisions. We can make wrong calls for different components. Yet previous failures can develop design acumen for Big Data platforms. A Big Data platform should adapt to changes to the organization, advances in technology, and new requirements. In many cases, decisions we make have different trade-offs. Each trade-off depends on the circumstances. When designing Big Data platforms, we should understand the trade-offs we are making. It helps to conceive the requirements and what we can do about them better.

Designing a Big Data platform requires craftsmanship efforts from many directions. There are always constraints when crafting the software. These constraints often result in trade-offs. Each trade-off introduces challenges in different directions. Some of these challenges in Big Data platforms are discussed in the previous chapter. As platform engineers, we need to outline the challenges concerning the trade-offs we make. Furthermore, some of the trade-offs when designing Big Data platforms will be now gone over.

12.9.1   Quality vs Efficiency

While the amount of data increases for a particular job, the quality usually improves. Scanning the whole data set can give better quality, but is it worth doing so? Full processing can take time and resources. In many circumstances, quality improvement might become negligible after a point. If the organization can sacrifice from quality a bit, then it is possible to process fewer data. Sampled data generally implies less resource usage and improved delivery times. Big Data platforms should allow tuning of the data via sampling. It is up to data customers to decide how much they want to consume. Platform engineers should think about the sampling part of the design and how to enable it.

12.9.2   Real time vs Offline

Online processing processes the data as it comes and can give quick updates. On the flip side, offline processing processes data after accumulating data for a bigger window and can give more accurate responses. Engineers often have to trade off accuracy for responsiveness or vice versa. For some organizations, accuracy is an absolute must, and response time is not as important. In these cases, it is possible to go for an offline system. For others, responsiveness is an important constraint. So, a stream processing solution fits better. In some cases, organizations have to employ both. While designing a Big Data platform, engineers should choose a system depending on the online/offline requirements and might need to tweak the platform later on.

12.9.3   Performance vs Cost

Organizations want to make individuals productive when querying, exploring, and summarizing data. If users wait too long to get their results back, it ends up in a loss of time. To get better response time, organizations might want to scale their computing infrastructure horizontally. Adding more nodes to Big Data systems improves the responsiveness to a certain degree. Nevertheless, adding more nodes to an existing system comes with additional cost. The platform engineers want to make users happy but keep the running cost of systems under control. Performance tests against systems can unveil performance characteristics. The bottleneck may happen due to contention during working hours or heavy ETL jobs. With data at hand, engineers can decide how to invest in a different set of systems and optimize user experience concerning the cost of investment.

12.9.4   Consistency vs Availability

Big Data platforms consist of connected distributed systems. In the realm of distributed systems, platform engineers always have to decide consistency vs availability. In general, consistency means that reads from a database are up to date. Clients querying the data gets the same view of the data. Availability means that queries against a database always return a response. Note that the definition given here is relaxed as opposed to the CAP theorem (Brewer, 2000) where it uses very strict definitions (Kleppmann, 2017). In practice, most Big Data systems do not really conform to the actual CAP theorem. Most systems rather have a more relaxed approach to availability and consistency. Even with the relaxed approach, it is not so easy to satisfy both availability and consistency when there are multiple nodes involved. For some organizations, consistency might be important due to the nature of the processing. For some others, response time or getting a result is more important. Platform engineers often have to evaluate the different systems and relax either consistency or availability requirements.

12.9.5   Disruption vs Reliability

Big Data platforms contain many different systems underneath. Each system has different fault tolerance and resiliency characteristics. While we want all systems to be as reliable as possible, it is often hard to do so for a wide range of systems. Hence, some systems occasionally may go down or become unresponsive. For disruptive events, platform engineers need to decide whether to engage an engineer to fix the problem or not. Obviously, the more critical the system is, the more aggressive the team has to deal with disruptions to make it more reliable. Nevertheless, some systems do not need to be as reliable. The system can be down outside of business hours and still be alright. The trade-off is to define how aggressively platform engineers should engage with a different set of systems. With a limited amount of engineers, the platform engineers have to sort systems from most critical to the least. After defining priorities, the team can decide how much disruption is acceptable for each system and set up alerting accordingly.

12.9.6   Build vs Buy

Organizations often need to define what to build vs buy for their Big Data needs. One can use Presto for ad hoc querying. Yet, it requires setting up a cluster and maintaining it. On the flip side, one can use a cloud solution for ad hoc querying and warehousing. The same sort of decision goes to many different components ranging from discovery, messaging, pipelining, and so forth. Moreover, there might be times where none of the options suffice the needs of infrastructure. In those cases, the organization has to invest in the development of a brand new system. Build vs buy decisions are quite hard. It involves many different aspects, such as talent, cost, hiring, infrastructure, and so forth. Platform engineers can make some of these decisions. In many cases, executive sponsorship or approval might be needed for such decisions. Platform engineers can prepare a clear compare and contrast documentation to choose the right solution to the best of their effort.

12.10 Data Ethics

Many aspects of Big Data platforms have been discussed, but data ethics is tackled last. Data ethics is about moral responsibility toward generation, recording, curation, processing, consolidation, sharing, and the use of data. Although Big Data platforms provide immense value to organizations, they should not violate the rights of individuals. As engineers building such platforms, we should put the rights of individuals over corporate profit or benefit. Human rights and data protection laws should be adhered when making decisions about Big Data platforms.

Data ethics is a way to look at data from an emphatic perspective. People's data should be treated as if they belong to us or someone close. Domestic laws may cover a good part of personal data handling. Nonetheless, it is found that an emphatic approach is easier to follow and apply. In the lights of empathy, guiding principles can be followed DataEthics (2017):

- Human rights should be at the center of data processing.
- Humans should be in control of their data and should be empowered by data.
- Data processing activities should be transparent to users. All activities should be explainable in simple language.
- The data processing organization should be accountable for their activities. Both internal members and subcontractors should be morally responsible for their operations.
- Data processing should not be used to discriminate against individuals. User profiling should not end up working against people due to their income status and social position.

Appendix A

Further Systems and Patterns

Throughout the book, we have touched on many subjects. Some of the subjects would have been great to add but might not be appropriate with the flow of the book. Thus, I have moved these subjects to the appendix to give a rough idea of them. In this part, I would discuss Lambda architecture, Apache Cassandra, and Apache Beam.

A.1   Lambda Architecture

Lambda architecture is a deployment model where organizations complement batch processing with stream processing for real-time big data problems. It has arisen due to troubles in serving data in real-time (Marz, 2011). Ideally, a system wants to scan entire data to respond to a query. In practice, responding to a query gets tricky since there is just so much data to scan for some queries. The data volume can result in outrageous response times. Moreover, organizations choose availability over consistency. Most organizations would prefer services to be available. Choosing availability over inconsistency results in weaker consistency levels. A read after write might not return the expected response. Without read repairs, the data can stay corrupted. Human error can also lead to problems. Updates to systems pose corruption threats that cannot be recoverable (Figure A.1).

To address these problems, the Lambda architecture uses an immutable stream of data and leverages precomputed batch views in a combination of stream computing. Storing data immutable eliminates the chances of data corruption. Nevertheless, immutable data leads to a more complicated computation. To compute queries, Lambda architecture has a batch layer, speed layer, and serving layer. The incoming data is pumped into both the batch and speed layers. The

Designing Big Data Platforms: How to Use, Deploy, and Maintain Big Data Systems,
First Edition. Yusuf Aytas.

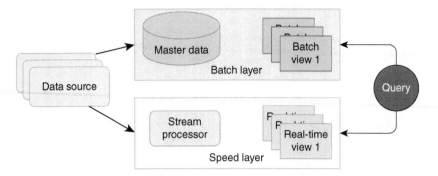

Figure A.1 Lambda architecture.

serving layer responds to queries by incorporating responses from the batch and speed layer.

A.1.1 Batch Layer

The batch layer accepts an immutable record stream of data and persists it to durable storage like Hadoop distributed file system (HDFS). It keeps the master data. Any state updates can get recomputed through the batch layer. The batch layer runs batch tasks to generate views for the serving layer. The batch layer generally lags quite some time compared with the speed layer. It can potentially fix the problems in the streaming layer as it has the full set of data. The batch layer recomputes the views periodically and outputs the result to a read-only store.

A.1.2 Speed Layer

The speed layer is responsible for computing real-time view on the fly. It accepts data from one or more data sources and generates real-time views from the streaming data. Once it computes the views, it saves the output to a data store that can naturally accept random writes. Typically, the speed layer uses a stream processor, such as Apache Flink or Apache Storm. Redis and Apache Cassandra are well-suited storage solutions for real-time views.

A.1.3 Serving Layer

The serving layer handles the merging of real-time and batch views. It fetches the latest views from both speed and batch storage and combines results to generate an aggregated view. The key for the serving layer is to minimize the response time for a given query. Hence, the serving layer can parallelize queries for the views. It can then generate an aggregate view when both results are complete.

A.2   Apache Cassandra

Apache Cassandra is a proven distributed NoSQL database. Many organizations have adopted Cassandra as their NoSQL database choice. Cassandra provides high availability, linear scalability, faulttolerance, and multiregion replication. It replicates the data automatically between different nodes as well as between data centers. The replication mechanism provides fault-tolerance and surviving data center outages. There is no single point of failure at any time. Applications can run smoothly even if a node dies or joins to the cluster. With new nodes joining the cluster, Cassandra can store massive amounts of data (Cassandra, 2020). Thus, it becomes a good choice to persist data permanently for big data platforms.

Cassandra comes with its own query language called CQL (Cassandra Query Language). Nevertheless, the syntax is quite similar to the traditional SQL. Cassandra is a key-value store. It neither has joins nor indexes. It provides some filtering functionality; however, it is not practical as it would hurt performance. Cassandra also does not have transactions. The application has to embrace failures and handle failures itself rather than relying on Cassandra. Failure handling should be part of the application logic as Cassandra provides tunable consistency. With weaker consistency levels, it may not be possible to read the data after a write.

A.2.1   Cassandra Data Modeling

Designing data models in Cassandra requires some mental adjustment. Cassandra users have to think about how to query data while modeling. It has concepts of the partition key and clustering key that form a primary key. Queries should ideally hit one partition and read all relevant data. The partition keys are responsible for data distribution across Cassandra nodes. The partition keys are used to locate and retrieve data. Clustering keys are responsible for data sorting within the partition. Multiple columns create a composite key to lookup partitions. Let us see an example Cassandra table where we would store user interactions:

```
CREATE KEYSPACE dbdp WITH REPLICATION =
    {'class': 'NetworkTopologyStrategy',
        'data-center-1': '5',
        'data-center-2': '5',
        'data-center-3': '3'}
    AND durable_writes = true;

CREATE TABLE dbdp.user_interaction (
    user_id             uuid,
    interaction_type    text,
    created_at          timestamp,
    interaction_values  text,
    PRIMARY KEY ((user_id, interaction_type),created_at)
```

```
) WITH CLUSTERING ORDER BY (created_at DESC)
  AND BLOOM_FILTER_FP_CHANCE = 0.1
  AND CACHING = {'keys': 'NONE', 'rows_per_partition': 'NONE'}
  AND COMMENT = 'A table to store user interactions'
  AND COMPACTION =
    {'class': 'org.apache.cassandra.db.compaction.
     TimeWindowCompactionStrategy',
        'compaction_window_size': '3',
        'compaction_window_unit': 'DAYS',
        'max_threshold': '32',
        'min_threshold': '4'}
  AND COMPRESSION =
    {'chunk_length_in_kb': '4',
        'class': 'org.apache.cassandra.io.compress.LZ4Compressor'}
  AND CRC_CHECK_CHANCE = 1.0
  AND DCLOCAL_READ_REPAIR_CHANCE = 0.0
  AND DEFAULT_TIME_TO_LIVE = 86400
  AND GC_GRACE_SECONDS = 259200
  AND MAX_INDEX_INTERVAL = 2048
  AND MEMTABLE_FLUSH_PERIOD_IN_MS = 0
  AND MIN_INDEX_INTERVAL = 128
  AND READ_REPAIR_CHANCE = 0.0
  AND SPECULATIVE_RETRY = '7.0ms';
```

In the example model, there are just so many things going on. Cassandra gives so much flexibility while defining tables. We can tune tables in many ways to get optimum performance. We create a keyspace with a multiregion replication setup in the first statement. We can define different replication factors (RF) for each data center. In the following statement, we create a table to store user interactions. The table has user interaction type and user id as partition keys to distribute data over nodes. We also defined a clustering key to store creation time and sort data within partition according to creation time. The rest of the table is to store interaction specific values. Apart from the column definition, the table has so many properties. Some of the notable properties are compaction, compression, and so forth. We will touch on some of these mechanisms later. Let us see how we can retrieve data:

```
SELECT interaction_values, created_at,
    WRITETIME(interaction_values) AS wrote_at,
    TTL(interaction_values)       AS ttl
FROM dbdp.user_interaction
WHERE user_id = 4c4e67a4-1453-11eb-adc1-0242ac120002
      AND interaction_type = 'PAGE_VIEW'
      AND created_at > '2020-10-21 11:43:02.000'
```

We have to supply all partition keys while querying because Cassandra has to locate the data by looking at partitions. Cassandra stores TTL (time to live) and write time per column. We can retrieve them with the helper functions in the example. One important thing to remember is to make Cassandra queries idempotent for both writes and reads so that it can retry queries in case of failures.

Basically, we should supply all values directly and set idempotent to true. More-over, it is possible that we would have to denormalize data and select columns depending on the needs since there are no joins. Cassandra also does not support multi-partition retrieval. When we need multiple partitions, we should query them separately in parallel where it is possible.

A.2.2 Cassandra Architecture

Cassandra architecture allows it to handle reads and writes across many nodes without a single point of failure. It distributes data among nodes with partitions. Every node is equal. Each node gets its share of data. In case of failures, other Cassandra nodes can cover and redistribute data over remaining nodes. Cassandra clients keep a list of nodes to retry operations when a failure occurs. They discover Cassandra nodes through the node supplied in the connection. Clients write and read through a coordinator node. Any node in the cluster can be a coordinator for a given operation. Coordinator coordinates operations among the replicas for the data based on replication factor.

Cassandra depends on many different ideas that one needs to understand to make the best use of it. Some of these concepts will be gone through, such as partitioning, virtual nodes, multi-master replication, and cluster membership. Later, we will discuss the storage engine concepts, such as CommitLog, Memtable, SSTable, compaction, repairs, and so forth (Figure A.2).

A.2.2.1 Cassandra Components
Cassandra relies on techniques like request partitioning over data set, node mem-bership, automatic failure handling, incremental scaling, and so on.

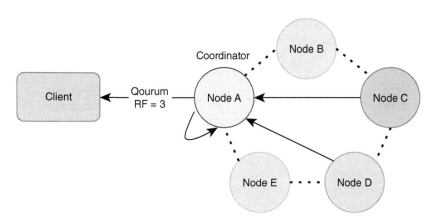

Figure A.2 Cassandra architecture.

Data Set Partitioning   Cassandra achieves horizontal scaling by partitioning data over many nodes and uses the last write wins approach to resolve conflicts in the case of conflicts on a partition. When partitioning data, Cassandra uses a technique called consistent hashing (Karger et al., 1999). Consistent hashing allows Cassandra to distribute data over nodes independent of the node number. The main advantage of consistent hashing is about redistribution of data when a node fails or gets added to the cluster. Consistent hashing requires only a fraction of data to move.

Virtual Nodes   If there are many nodes in the cluster, the chances of having an imbalance between nodes are less. If there are not many physical nodes, the chance of imbalance is higher. To address this downside, Cassandra uses so-called virtual nodes (vnode) to distributed data more evenly across nodes. Virtual nodes help to balance query load and disruptions easier.

Multi-master Replication   Cassandra replicates data across many nodes depending on the replication factor. The replication factor defines how many times data needs to be replicated among Cassandra nodes. When a mutation occurs, the coordinator node hashes partition key to find the nodes that data belongs to. Cassandra offers different replications strategies to allow different setups like multiregion replication. Moreover, Cassandra uses a mechanism called snitch to spread replicas around the cluster to avoid correlated failures like rack failures. If the number of racks is greater or equal to the number of replication factors, Cassandra chooses replicas from different racks for maximum durability.

Tunable Consistency   Cassandra allows tuning consistency per operation by choosing between availability and consistency. It comes with a menu of consistency levels, such as *LOCAL_ONE*, *LOCAL_QUORUM*, *ALL*, *ANY*, and so on. For *LOCAL_ONE* consistency, only one replica needs to respond from the local data center. For *LOCAL_QUORUM*, a quorum of replicas need to respond from the local data center. A quorum is the majority of replicas for a given replication factor, for example, two nodes for replication factor 3 and three nodes for replication factor 5. Note that writes always sent to other replicas regardless of the consistency level. Cassandra waits for enough nodes to respond before returning to the client.

Cluster Membership   Cassandra uses a gossip protocol to broadcast membership information between nodes. Cassandra nodes exchange information about both themselves and other nodes that they are aware of. Every node in the Cassandra cluster talks to other nodes periodically. Gossip protocol tells the basic membership information. Nodes also need to know whether a node is up or down.

They determine it through a failure detector mechanism based on the heartbeat rate. In case of failures, the Cassandra nodes store writes locally and waits for a period for the failed node to wake up. If the node wakes up, Cassandra nodes send updates to the node. This mechanism is called hinted handoff.

A.2.2.2 Storage Engine
Cassandra storage engine uses many mechanisms to make it performant and reliable at the same time. We will go through some of these mechanisms.

CommitLog CommitLogs are local append-only structures for a node. Any data written first gets written to the commit log. Commit logs provide failure recovery in the case of unexpected crashes or shutdowns. On startup, Cassandra will read the commit log and apply mutations. All mutations are write-optimized. Cassandra keeps commit logs by segments and flushes commit log segments once it is safe to do so.

Memtable Memtables are in-memory structures that Cassandra buffers write. Cassandra keeps one Memtable per physical table. Memtables eventually become SSTables once Cassandra flushes them into the disk. Cassandra flushes Memtables depending on the cleanup threshold or commit log size.

SSTable SSTables are immutable data structures Cassandra writes to disk. Cassandra runs compaction tasks where it merges one or more SSTables into one. Once Cassandra creates a new SSTable, it can delete the older one. Cassandra offers different compaction strategies to optimize write or read performance depending on the needs. It can also decrease SSTable size by compression.

Repairs The nature of preferring availability over consistency results in inconsistencies between nodes, which might result in data loss when the node gets replaced or tombstones (markers for deletes) expire. Cassandra fixes these inconsistencies by running repair jobs. Repair jobs synchronize nodes by comparing them with each other and stream the difference between them.

A.3 Apache Beam

Apache Beam provides a unified model for stream and batch processing. Apache Beam pipelines can run in different environments. Apache Beam supports extensible APIs and SDKs to write new connectors, libraries, and runtimes. It also supports client SDKs in many languages, which includes Java and Python. The beam comes with support to many backends, such as Apache Flink, Apache

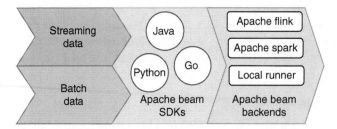

Figure A.3   Apache Beam programming model.

Samza, and Apache Spark. Using beam SDKs, we can represent data operations on any data size regardless of boundness. Beam uses the same classes to apply transformations for both bounded and unbounded data (Beam, 2020) (Figure A.3).

Apache Beam SDKs provide the functionality to build a data pipeline. In the simplest form, SDK provides functionality to retrieve data from different IO resources, transformation utilities to transform data into the required format, and lastly, exporting functionality to transfer data to the backend. Users can run the created data pipeline on the runners. Runners are distributed data processing systems to parallelize the computation. Let us now see the programming concepts for Apache Beam.

A.3.1   Programming Overview

Apache Beam accepts a driver program that defines pipeline with the help of Beam SDK. The driver program describes inputs, transformations, and outputs for the pipeline. The driver program also sets the execution options for the pipeline. The Beam SDK provides abstractions to define complex distributed data processing in a pipeline. The same abstractions work for both batch and stream processing. The primary abstraction is a pipeline that defines data processing from start to end. Pipelines use PCollections, which represent a distributed data set. PCollections can be bounded and unbounded and mostly come from an external data set; however, we can create in-memory PCollections as well. PTransforms represent the steps in the pipeline. Every PTransform takes one or more PCollections and produces zero or more PCollections. Let us now see an example:

```
final PipelineOptions options = PipelineOptionsFactory.create();
final Pipeline pipeline = Pipeline.create(options);
pipeline
    .apply("(1) Read all lines", TextIO.read().from(inputFilePath))
    .apply("(2) Flatmap to a list of words", FlatMapElements.
    into(TypeDescriptors.strings())
        .via(line -> Arrays.asList(line.split("\\s"))))
    .apply("(3) Remove all non-alpha-numeric", MapElements.
    into(TypeDescriptors.strings())
```

```
        .via(word -> word.trim().toLowerCase().
        replaceAll("[^a-z0-9]", "")))
    .apply("(4) Filter empty strings",
    Filter.by(input -> !input.isEmpty()))
    .apply("(5) Count words", Count.perElement())
    .apply("(6) Sum counts", Sum.longsPerKey())
    .apply("(7) Top 10", Top.of(10, new KVComparator()))
    .apply("(8) Convert KV to string", FlatMapElements.
    into(TypeDescriptors.strings())
        .via(kvs -> kvs.stream().map(count -> count.getKey() + " --> " +
        count.getValue())
            .collect(Collectors.toList())))
    .apply("(9) Write output to a file", TextIO.write().
    to(outputFilePath));
//Run pipeline until it finishes
pipeline.run().waitUntilFinish();
```

In the example, we read words from an input location. It can be both remote and local locations. We then parse it line by line and find the count per word. We sum the counts and then find the top 10 most used words. Later, we output the words to the output directory specified. Once the pipeline is ready, then we can run it locally. We can also run the pipeline on various backends, but we have to specify the runner in the pipeline options.

PCollections look like a regular collection, but they have some important features as follows:

- PCollections only belong to a pipeline they are created in.
- PCollections are immutable.
- PCollections have a schema, such as Protocol Buffers, Avro, and JSON.
- PCollections do not support random access.
- PCollections do not have any bound. They can get big as needed.
- Each element in PCollections has a timestamp.

PTransforms get applied to one or more PCollections. Each PTransform has a generic apply method. Each transform operation returns a PCollection. PTransforms create a directed acyclic graph (DAG) where PTransforms are functions that accept PCollection nodes as inputs and emit PCollection nodes as outputs. We can chain PTransforms to come up with a complete pipeline like the example. A typical pipeline in the simplest form looks like as follows (Figure A.4).

Figure A.4 Apache Beam pipeline.

A.3.2 Execution Model

Apache Beam employs different runners. The execution characteristics depend on the runner. Runners serialize elements and communicate with each other between transforms and use serialization and communication to route elements for grouping or redistribution. On the flip side, some transforms can happen in the same worker. In these cases, the runner would pass data to the next transform. Between the transforms, the runner might use checkpointing to persist the result of a transform operation.

Beam pipelines attack parallelism problem. Thus, some operations that require sequencing elements are not easily expressible. Instead of processing elements one by one, Beam processes elements of PCollection in bundles. A bundle is a bag of elements specified by the runner. Having bundles allows runners to save intermediary results when necessary. Runners' control over bundle size also enables optimizations depending on the batch/stream processing. Batch processing might use a bigger bundle size than stream processing.

ParDo is the core element of a transform function. When executing a ParDo, Apache Beam divides the input elements into bundles. The maximum parallelism is equal to the size of PCollection where bundle size is equal to 1. If the runner decides to run ParDo back to back on a given bundle, then it can create dependently parallel ParDo transforms illustrated in Figure A.5. Executing dependently parallel ParDos avoid redistribution of data and potential serialization/communication costs.

If a failure occurs during the processing of a bundle, the entire bundle fails. The bundle needs to be retried. If the bundle still fails, the entire pipeline fails. If there are dependently parallel ParDos, then a failure in any of the bundles would fail all ParDos. All ParDos in the chain needs to be retried. This effect is called cofailing. If the retry does not succeed, the pipeline operation fails.

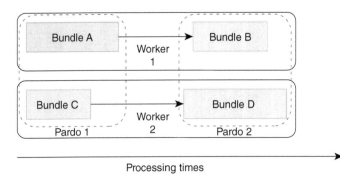

Figure A.5 Apache Beam ParDo processing.

Appendix B

Recipes

B.1   Activity Tracking Recipe

B.1.1   Problem Statement

Activity tracking is an indispensable component for organizations. It provides the ability to learn about users and brings a personalized experience to them. User activity tracking can help to optimize key performance indicator (KPI) for the organization. Activity tracking involves capture user activities while the user is browsing the website or mobile app. Each activity results in one or more signals that are pushed down to the big data infrastructure. The response to these signals can be computed as they come or computed through offline processing. The response can be as simple as a score or a Boolean value that determines what the user experiences.

B.1.2   Design Approach

Computing different scores and rates for the user might require a mix of stream and batch processing. Some values can be computed on the fly. Others might require extensive batch pipelines. Considering both requirements, we divide the problem into two subproblems. The first problem is the data ingestion and persistence. The second problem is computation.

B.1.2.1   Data Ingestion
The data ingestion should support both streaming and batch processing needs. The signals should be fed into persistent storage with appropriate retention. Our proposed approach uses a messaging layer component, Kafka, to store signals for a certain amount of time in the messaging layer. At the same time, Kafka feeds object storage for long-term persistence. Each signal has a well-defined schema and gets registered to the data discovery tool, Amundsen. We employ a data ingestion service that exposes an API that is compatible with OpenAPI specification.

Designing Big Data Platforms: How to Use, Deploy, and Maintain Big Data Systems,
First Edition. Yusuf Aytas.
© 2021 John Wiley & Sons, Inc. Published 2021 by John Wiley & Sons, Inc.

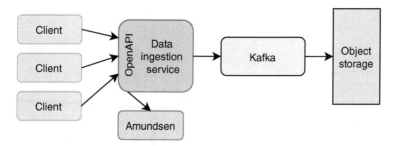

Figure B.1 Data ingestion components.

Amundsen parses API definitions and stores it internally for data discovery purposes.

As it is depicted in Figure B.1, the data ingestion service uses OpenAPI specification for external clients to publish new user signals. Each new user signal gets registered to Amundsen for data discovery purposes. Data ingestion service is stateless. It only facilitates publishing signals to Kafka. The data ingestion can scale horizontally as needed. On the Kafka end, each user signal can get its own topic. Each topic can have a different number of shards depending on the signal volume. Kafka also integrates with object storage through a Kafka sink. It can partition data based on time through a partitioner.

B.1.2.2 Computation

Computing scores out of user signals might require many different configurations. Some of the scores might be computed easily through stream processing. On the flip side, others might require some longer time periods. Since the data is available both in Kafka and object storage, it is possible to make use of stream and batch processing. Typically, computation results need to be served pretty quickly. Therefore, it is not feasible to compute the results over and over again. Caching results would solve the problem. Stream processing can directly write to the cache. Batch processing can write results to intermediary tables and load the data to cache. The results can be recomputed by leveraging Kafka or object storage.

As can be shown in Figure B.2, the computation tasks involve Apache Spark to address both batch and stream processing needs. Spark streaming receives data directly from Kafka and applies updates to Redis. Spark streaming can utilize different time windows where the records in a window can be comfortably stored. On the batch computation part, Spark utilizes both object storage and other data sources to compute the expected score or rate. The results can be potentially saved to an intermediary table so that it is possible to retrieve them if Redis goes down. In the last part, services consume the computed scores or rates from Redis to determine various personalized behaviors for the application.

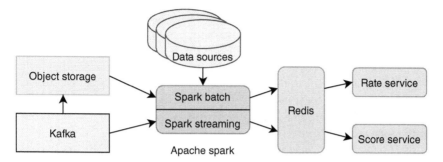

Figure B.2 Computation components.

B.2 Data Quality Assurance

B.2.1 Problem Statement

When designing a big data process, we expect the process to produce software that adheres to certain quality standards. Even though we can deterministically look for certain tried and true quality standards in our big data products, the ultimate judge of this software will be the consumers of big data. Consumers can be actual users or internal services and/or data. In the scope of this recipe, we will skip end-product quality here. Rather, we are going to talk about how to design a process that will enable you to look for the aforementioned quality aspects continuously. In a sense, we are going to look into how to create a cycle that lets you create "Quality Software and Actionable Data." After all, you cannot expect quality data products from a lesser-quality process.

B.2.2 Design Approach

Our approach to quality assurance has worked quite well for several sectors throughout the years. It has enabled many well-known global businesses to produce quality software rapidly and incrementally. It involves ingredients and preparation to bring quality to big data applications.

B.2.2.1 Ingredients

Applying quality assurance requires the involvement of the entire organization. Every part of the organization is also part of the ingredients.

Cross-Functional Agile Team Teams should embrace iterative development and have some knowledge of agile methods. Nevertheless, it is not necessary to go with certain frameworks such as Scrum or Kanban. The key is to establish reasonable iterative process and ability to adjust to new requirements.

Nonhierarchical team members: While the team members, especially the data product owner (DPO) might have different responsibilities, it is advised to give all team members an equal say in all matters. Obviously, expertise is important. Team members will gravitate toward certain ideas voiced by the person(s) who are experts in those areas, but that should not prevent the team from discussing most of the things together. Everybody should feel comfortable voicing their opinions. No question should be considered as a stupid question in the team.

Dedicated roles: Following roles should be covered:

DPO (data product owner): A person who knows what is to be built, has the authority and aptitude to make decisions on the requirements, and is the main bridge between the team and the data customer.

BA/SME (business analyst/subject matter expert): Someone who owns the backlog and epics/stories in the backlog, helps the team to clear out the confusion or conflicts in the backlog, and champions proven standards in all backlog items and constantly works with the DPO and the customers to keep the team working on valuable stories. In a big data project, this person might very well be a data scientist.

Data engineer: The person who creates working software or data pipelines and champions good code quality standards and performs test-driven development (TDD) with the other developers they pair with.

QA: The one who asks questions. Quality assurances (QA's) main responsibility should be to make the team aware of what might go wrong on the way. They champion the team's responsibility for the quality, making the team aware of what type of test needs to be done by whom and also when. For instance, in a data analysis project, the main job of QA is to scrutinize the findings and see if they are in line with spot checks and small sample sets. Once it looks alright, we can automate most of the tests on the data to make them repeatable and easily reportable.

Data platform engineer: The person who helps the team to build a robust continues integration (CI)/continues deployment (CD) pipeline, is involved in key decisions in designing an overall big data platform, and knows the DevOps methods and drives the team to achieve good practices. They also drive the team to use the best tooling available at the time to achieve the best results.

Executive Sponsorship Sponsor is a person or group of people who support the team to self-manage their work and help people to set their own pace and let them produce their best work promptly while consulting with the DPO to deliver the best solution for the business.

Customer Involvement They are the people who will actually use your software or data that are being produced and who knows the ins and outs of the business and can guide the DPO and the team to define the problems the software is supposed to help solve.

Self-Sufficient Infrastructure The team can build all necessary environments and/or utilities that will enable them to produce the best software possible. For instance, the team might need a Kubernetes namespace for the team. Once it is set up, they control the whole deployment aspects of new data sets and frameworks. With this approach, there is no wholesome bureaucracy for one single deployment.

Dedicated Team Space Assuming that the COVID-19 pandemic is mostly over and people can physically work together again, the ability to raise your head and ask a question to your teammate is vital. A few minutes of side-to-side pairing is usually much better than hours of online back and forth over documents or documentation systems.

Communication and Monitoring An ideal team space would be littered with whiteboards, roadmaps full of pictures and stickies, monitors showing build statuses and metrics from production, etc. These enable the team and stakeholders to keep an eye on the goals, make informed decisions, and focus on what to do next.

B.2.2.2 Preparation

The preparation has the following items:

- Find out what you want to build as a team and involve all team members. Inception is an activity where the team members get together with the customers or in a data project, consumers of data, to come up with what to build (Caroli, 2017). It requires a lot of investigation works and has activities that help determine what is actually needed.
- Have your DPOs, BAs, and QAs create and groom the backlog all the time. One of the most important features of good quality software is its ability to adapt. Your data-oriented people should be looking at your stories constantly to keep them up to date with the ever-changing requirements. Your QAs should be feeding into the process by reflecting on the previously discovered problems like bugs or unexpected results from inadequate data. In short, it is about getting the feedback from the previous iterations and feeding them into the remaining backlog items.
- Have your data engineers create a robust and reliable CI/CD pipeline. The only way to get the most valuable feedback is to have the ability to test and deploy to your production system (or your data consumers) safely and rapidly. For data-specific scenarios, your team should have the ability to setup/test/teardown rapidly. Your data pipelines must be fed fresh data continuously; all tests on that data need to be performed quickly (preferably automatically most of the

time), and the resulting output should be deployed to production to get the real feedback from consumers rapidly.

- Have your Devs perform TDD and pair them with your QAs a lot. Without getting into details of TDD, it is heavily advised to have your developers move into a TDD mindset. It is not about finding bugs and testing specifically, but it is more about pushing them to think about the problem from the get-go. Write something that they want to accomplish, see it fail, write the code to overcome that problem, and finally look for ways to do it more efficiently: red/green/refactor. While the Devs are thinking about what kind of tests to cover, it is very beneficial to let them pair with your QAs to have the inquisitive minds of your testers to help your developers see alternative ways of things going wrong.

- Perform three amigos. Three amigos are business analysts, the developers, and the QAs. They get together when they are picking a story for a few minutes to agree on what the story is (story kickoff). They get together again when the story implementation is complete to have a quick look at what the story looks like when it is implemented (story desk check).

- Make sure your roles have access to DPO and BA constantly. The team needs to ask business (or data)-related questions to your SMEs immediately when they arise. Since your DPOs and BAs are the maintainers of your backlog, they have the most up-to-date vision for the product or the service. They should be guiding the team constantly.

- Get a lot of feedback regularly from the consumers of your service or data. Nothing fails like Chinese whispers. Do not let your team assume what the customer actually wants. Always create healthy communication with the people who will acquire your services so that you actually build something that solves a real problem.

- Deploy working software to production frequently and be ready to fix things quickly. Do test in production (blue/green deployments, zero downtime deployments). Regardless of what you're delivering, deliver it fast to the consumers, and get their feedback rapidly. Your best feedback will come when your product or data is actually used. It is also important to be able to tweak and test things in production to see what the effects are. As long as your mean time to recover is very low, you should not be afraid to experiment on production. Long gone are the days where the prod is untouchable, and everything must be caught before going to prod.

- Have your QAs test and scrutinize the findings. Let them automate tests to be informed quickly. Make use of good reporting and monitoring to be informed and alter your previous decisions. Let them do automated spot checks on your software/data, and make that automation smart enough to test different things every time. Use small, randomly selected sample data sets to test assumptions, have real people look at the data for not easily discoverable anomalies, and have certain fixed tests continuously testing for little things to catch possible differences between versions. The examples can be much more. It all depends on the context. The basic idea is to have some people who like to think about what can go wrong and have a look at your product constantly.

- Implement a well-structured Testing Pyramid. Know what to test, how to test, and whom to have it performed. In a data project, we can assume there will be unit tests that will test the data quality while ingesting it. There would be integration tests that examine if the intermediate data stores are there and able to be written on/read from. There would be functional tests that examine data integrity. There would be journey tests where some set of data is introduced to the source system and expected to come out of the process in a certain way. The Test Pyramid is about having more and more tests where the scope is smaller and reducing the number of tests as you increase the scope of them (Vocke, 2018).

B.2.2.3   The Menu

As you can see, we have talked very little about QA-specific topics or testing itself. The reason for that is simple; for a good-quality approach, you need to change the ways of working of the team. It is not only about adding more automation scripts or more testing into the mix. The team owns quality. However, it is about setting up a process where there are different sets of eyes looking at the problem from different angles. Also, it is about decreasing the feedback loop between the users of the service/data and the team who builds it. We will never be able to fix all the bugs or remove all the problems that software might produce. Our main objective should be to maximize the collaboration of all parties who want the software to exist. It should be to provide safe and fast ways to experiment, try new things, and fix them quickly if they break things.

Alper Mermer put a huge effort into making this recipe real. I have been fortunate enough to collaborate on this recipe with him.

B.3   Estimating Time to Delivery

This recipe covers the deployment of machine learning models within a stream processing system to provide near-real-time predictions based upon a combination of data that is provided within the streaming data itself as well as data from other external data sources.

B.3.1 Problem Definition

Consider the scenario where we are developing a food delivery service that allows customers to order food from their favorite restaurants and have it delivered to them. Every time a customer places an order from a restaurant, it is ideal to provide the customer with an estimated delivery time. We can implement time to delivery by leveraging a machine learning model.

Typically, these machine learning models have been trained to take in a predefined set of inputs known as a feature vector and use that information to predict the delivery time. Every feature in the feature vector represents an individual piece of data that is used to generate the prediction. Examples of features for the time-to-delivery model include information from the food order itself (e.g. time of day, delivery location), as well as historical features. We can supply the model with features like average meal prep time for the last seven days and average delivery time in the city for the last one hour.

B.3.2 Design Approach

Models that have such stringent latency requirements cannot reliably expect to compute features that require access to traditional data stores in a performant manner. For instance, it is not feasible to directly query a database to compute the average meal prep time for a given restaurant over a specific period of time with a sub-second response time consistently.

Since we need to generate a prediction within a few hundred milliseconds and the feature vector for the time-estimation model depends upon data provided as part of the request, such constraints require a streaming solution. The core capability of any streaming system is its ability to continuously process endless streams of data and act upon it in near realtime.

B.3.2.1 Streaming Reference Architecture

Because there are so many streaming systems and everything is evolving so quickly, we have decided to narrow our focus to a representative composition of systems that define a reference architecture that provides all of the essential features of a streaming system. The components of the streaming architecture are depicted in Figure B.3. The diagram includes a messaging system along with zero or more data stores and computing engines. Inside each of the boxes, there are some of the more commonly used technologies that provide that capability. For instance, the core of the streaming architecture is the messaging system itself, which is responsible for ingesting, storing, and delivering the data to the computing engines for processing. The most common choices today for a messaging system are Apache Pulsar and Apache Kafka as indicated by the gray

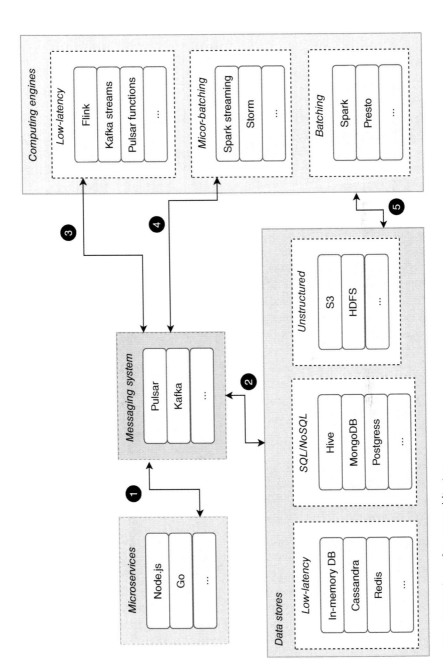

Figure B.3 Streaming reference architecture.

boxes. Let us walk through the architecture and dive into some of the details of the components and how they interact with one another:

1. Today's modern applications are increasingly being developed using a microservice architecture in which the entire application is broken down into small independent pieces of software that provide one particular service. The Node.js and Go ecosystems are both popular tools for developing these microservices. It is quite common to have these microservices utilize the messaging layer for inter-service communication.

2. Sometimes the raw data streaming into the messaging layer needs to be persisted directly to longer-term, persistent storage, and most messaging systems provide the ability to read/write data to and from various data storage technologies using custom connectors.

3. If our applications require low-latency (sub-second) processing on the data stream, then you can utilize one of the stream processing technologies shown to perform the necessary computations. It is critical that the processing engine has the ability to consume data directly from the messaging system and publish its results back with a minimal amount of overhead.

4. If our applications can tolerate longer latencies and/or an extra window of time is required to capture all of the data required to perform the calculation, such as calculating moving averages over longer periods of time, then a micro-batching engine is well suited for such computational tasks. As before, it is critical that the processing engine has the ability to consume data directly from the messaging system and publish its results back.

5. Traditional big data processing engines such as Spark or Hive are designed to consume data from long-term storage such as Hadoop distributed file system (HDFS) or object storage, and therefore batch processing jobs are unlikely to use the messaging system as an input source. However, this architecture also supports batch-oriented processing because it is often used to precompute certain values from historical data and cache it inside low-latency data stores, so the calculated results are accessible by the messaging system and low-latency processing engines.

Let us turn our attention back to the time-to-delivery estimation feature and discuss the implementation details of deploying the ML model and ensuring that we provide it with all of the data required in its feature vector. We will assume that the data science team has already developed and trained the model we want to use and specified all of the features that the model requires.

Therefore, our job is to deploy the model on a low-latency processing engine and develop a data pipeline that takes the customer order and retrieves all of the data necessary to populate the feature vector as shown in Figure B.4.

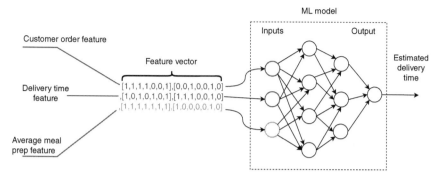

Figure B.4 Streaming feature vector. Given a trained ML model and feature vector definition, we need to take the customer order and generate a fully populated feature vector to feed to the ML to get an estimated delivery time for that order.

Adding to the complexity of this solution is the fact that the model needs additional data not available in the customer order itself and must request this data from other data sources. These often include a variety of disparate relational and non-relational databases. Unfortunately, models that need to provide predictions in near realtime cannot access data stored in such low-latency systems that can take seconds or more to respond. For instance, it is not possible to directly query the production database to compute the average meal prep time for a specific restaurant over a given period of time. Therefore, it is necessary to establish ancillary processes to precompute and index necessary model features. Once computed, these features should be stored in a low-latency datastore such as Cassandra where they can be accessed quickly at prediction time as shown in Figure B.5.

Let us follow the data as it flows through the architecture to produce an estimated delivery time for an incoming customer order:

1. The ancillary processes used to precompute and index necessary model features including the average meal prep time for the last seven days for every restaurant and the average delivery time in the city for the last one hour are scheduled to run periodically to cache the results in a low-latency data store, such as Cassandra, for sub-second retrieval when the order arrives. A batch-oriented processing engine is best suited for this job as it can process a large volume of data efficiently. For instance, a distributed query engine such as Hive or Presto is well suited for querying a database to compute the average meal prep time for all restaurants over a specific period of time. Once these values are calculated, they are stored in a low-latency datastore for fast retrieval at runtime.
2. A microservice will publish the customer food order to a predefined topic inside the messaging system.

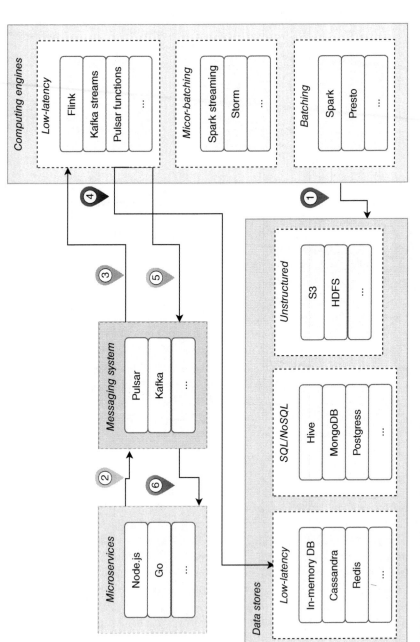

Figure B.5 The time-to-delivery estimation data flow.

3. A low-latency processing engine will consume the customer food order from the topic and begin the process of populating the feature vector. First, it will extract the features it can from the order itself, such as the time of day and delivery location.

4. Next, it will use other fields inside the order, such as the restaurant ID and delivery address to access the precomputed fields from the low-latency data stores. For instance, it will use the restaurant ID as the index into the Cassandra table that contains the precomputed "average meal prep time" values. Once the feature vector has been completely populated, the low-latency processing engine will send it to the ML model and get the predicted delivery time.

5. Finally, the low-latency processing engine will publish this result to a topic inside the messaging system.

6. A microservice will consume the message and display the value in the customer's mobile application.

David Kjerrumgaard put a huge effort into making this recipe real. I have been fortunate enough to collaborate on this recipe with him.

B.4 Incident Response Recipe

B.4.1 Problem Definition

When part of the data pipeline breaks, anomalies in important metrics should trigger an alarm. The alert goes to the engineer who is oncall to deal with such disruptions. On-call engineer is responsible to triage and mitigate problems as quickly as possible. What happens after the engineer engages with the problem can end up in many directions. Sometimes, the engineer might resolve the problem pretty quickly based on previous experience. On the other hand, the actions engineers take might lead to a catastrophe.

Handling disruptions is never easy and not all disruptions are the same. Responding to incidents requires preparing before the incident happens, keeping an event log, and sharing the findings through tickets and channels. The preparation involves minimizing disruptions and categorizing them. The rest is about finding an organized and practical way to deal with incidents.

B.4.2 Design Approach

Our experience in big data reveals that there are three tenets for incident response: minimizing disruptions, categorizing disruptions, and handling disruptions.

B.4.2.1 Minimizing Disruptions

Incidents happen. There's no way to stop them from happening. Nevertheless, we can attempt to make everyone's life easier. In big data processing, there are a couple of gotchas that we should be aware of. Obviously, deploying a new update on Friday is never a good idea. It is particularly problematic for big data as changes might reveal themselves a day later. What is more is that some deployments might not seem to break anything. They can be nasty. It might only affect a portion of data and might not be obvious. Since people do not check the data or reports over the weekend, the data slip through automated checks. The reality might become obvious in normal business hours after potentially costing revenue loss and additional processing needs. Hence, we should potentially avoid Friday deployments.

Another important aspect is to avoid cascading failures. Typical data pipelines might involve many teams. Some of them are downstream consumers of others. Some of the teams support the platform or infrastructure. A problem with infrastructure can delay or halt the processing of entire data pipelines. Engaging platform engineers in a timely manner is crucial. On the flip side, engaging the upstream data provider can prevent further delays and problems. Moreover, delays in multi-tenant environments might cause a significant backlog. Therefore, it is vital to define who owns what and engage owners, respectively.

B.4.2.2 Categorizing Disruptions

Categorization is an important but overlooked part of incident response. Engineers should be paged for absolutely right reasons. Having someone woke up for no good reason is absolutely the worst thing for employee retention and success. Asking the team to support half-baked pipelines or tools is just a setup for failure. Teams should avoid promoting solutions to production unless it is proven to be reliable. In the case of necessity, people behind the tool or pipeline should own the process until it stabilizes.

Having figured the right way to promote, the data team should categorize incidents into the following: requiring engagement at all times, requiring engagement in business hours, and requiring engagement before x time. Most of the paging software has a high and low paging setup. The team can use a mix of these to arrange the categorization. The key is to minimize the number of high priority pages. Avoiding paths that result in high priority pages should become a habit. High-priority paging should happen for unforeseen events. Anything that can be figured ahead of time should be part of the automation.

B.4.2.3 Handling Disruptions

The incident response consists of steps to resolve the issue as soon as possible while keeping a detailed log of events.

Triaging   Some metric has spiked or dropped. An engineer got a page. Perhaps, the number of rows associated with a table is much less than expected, or a heartbeat from a system was not seen for a while. The engineer has to assess the impact quickly. The engineer needs to evaluate how severe the problem is, who can potentially fix it, and which runbooks are available for the fix. The purpose of the triage is to determine if the problem can be addressed through available runbooks. If that is not possible, then engage the team or person who can help to mitigate the issue. Once the triage step is complete, the next thing is to communicate the findings.

Communication   The engineer has a rough plan of what to do at this step. She will either escalate it to the team or the person who might potentially fix it or follow runbook instructions. When the person or the team comes online, the on-call engineer needs to communicate the initial assessment and write down the actions she took. If the incident requires more than one team engagement, the on-call engineer should coordinate it. She can set up a call where anyone responsible can engage. A big part of the communication is the written event log. Engineers should log all actions and decisions step by step to the event log. Once all parties agree on a path to resolution, the mitigation step can take place.

Mitigation   Mitigation is about turning back system or systems to normal. It is not about fixing the problem, but the cause of the problem. The problem might be a recent config update, a column change, or an additional load. In the mitigation step, engineers might rollback the system or pipeline to a state where they know it works. The changes for mitigation are short term and rather workarounds. The issues should be properly fixed later on. The problem is though about processed data and downstream dependencies. A rollback procedure requires an extensive evaluation of the impact on downstream data. Affected pipelines or systems should be identified, put on pause, cleared off the changes, and restarted from the safe point. Nevertheless, it is always a very big challenge to organize such rollbacks as it requires vetting from many parties. Sometimes, it might be "ok" to roll with the erroneous data to resolve issues later on.

Retrospective   Once everything is back to normal, the team has to find out the root cause and make sure it never happens again. A common bad habit is not to take action about known incident types. Ignoring incidents result in the loss of time and missed opportunity for learning. Engineers should prepare a postmortem document to discuss things learned and what to do to avoid such incidents. A postmortem document is a great learning experience rather than putting a strain on the individuals. It is quite possible on-call engineer can make a mistake. It might be very late at night. The on-call person might be distracted or something else, especially if she was under fire continuously during oncall. Therefore, the postmortem should only focus on the environment and systems that made the incident happen. The retrospective is about making sure the root cause has been addressed in some way in the system or environment.

B.5 Leveraging Spark SQL Metrics

B.5.1 Problem Statement

Big Data is part of Industry 4.0. Its computation and storage requirements introduce the challenges according to traditional data processing approaches. To be able to deal with Big Data processing and storage challenges, data pipelines are widely developed for various use cases in the software industry such as social media, e-commerce, data analytics, finance, telecommunication, etc.

Apache Spark is one of the industry leaders in distributed computing engines and SQL on Hadoop solutions. It is also widely used for different use cases such as distributed data processing, stream processing, machine learning, interactive data analysis, etc. So, it is important to troubleshoot complex Spark queries and measure the production workloads in detail when using Spark. In this recipe, you will find the suggestions to build a new Spark SQL metrics pipeline and its benefits.

B.5.2 Design Approach

To get better information from Spark about metrics, we would setup a metrics pipeline and then consume it through database. We will now go through some of these steps.

B.5.2.1 Spark SQL Metrics Pipeline

Spark runtime metrics, such as application, job, stage, task, and executor, storage, are visible on SparkUI at the end of query execution. These metrics are also accessible by Spark public Rest endpoints. On the other hand, Spark-executed query metrics (aka SQL metrics) are also visible on SparkUI's SQL tab. Its Rest endpoint is going to be released with the next Spark release in addition to existing Rest endpoints.

SQL metrics are accumulators that are variables aggregating metrics data across the executors and can be written only by executors and readonly by the driver. On the other hand, SQL metrics are generated during query execution per physical operator. Capturing these metrics at the physical operator level is important for the following:

- Troubleshooting
- Understanding query complexity
- Physical operator runtime behaviors

- Performance analysis
- Measuring the production workload in a multi-tenant environment
- Long-term scalability planning

To leverage the above benefits and capture SQL metrics from the test, pre-rod, and prod environments per executed query, a custom Spark metrics pipeline can be built by consuming Spark Rest endpoints through Spark History Server for completed Spark queries, transforming nested JSON results to tabular format, and persisting a database. Then, SQL metrics can be queried and aggregated by tenant, application, physical operation, and/or date.

After Spark SQL metrics are persisted to the database, the following sample queries and aggregations can be run on the SQL metrics table in Figure B.6. Spark SQL metrics pipeline is illustrated in Figure B.7.

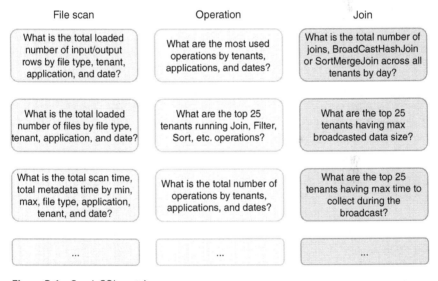

Figure B.6   Spark SQL metrics.

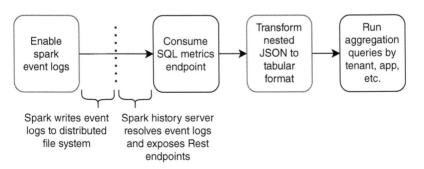

Figure B.7   Spark SQL metrics pipeline.

In addition to SQL metrics, the same metrics pipeline can be extended for other Spark runtime metrics, such as application, job, stage, task, executor, and storage to capture all exposed metrics and analyze:

```
case class Employee(id: String, name: String, department: String)
case class Address(id: String, address: String, city: String, country:
 String)

import spark.implicits._

val employeeDS = spark.createDataset(Seq[Employee](
   Employee("1","Employee_1","Department_1"),
   Employee("2","Employee_2","Department_2"),
   Employee("3","Employee_3","Department_3")))

val addressDS = spark.createDataset(Seq[Address](
   Address("1","Address_1","City_1", "Country_1"),
   Address("2","Address_2","City_2", "Country_2"),
   Address("3","Address_3","City_3", "Country_3")))

employeeDS.join(addressDS, Seq("id"))
   .filter(ds("id").>=(2)).sort(ds("id").desc)
```

After the sample query is executed, SQL metrics can be accessible by SQL metrics Rest endpoint. The SQL metrics Rest calls are simply *$sparkDriver*:4040/api/v1/applications/*$applicationId*/sql for running application and *$sparkHistoryServer*:18080/api/v1/applications/*$applicationId*/sql for finished applications. Some of the metrics we would get is as follows:

- Shuffle records written
- Shuffle write time
- Records read
- Local bytes read
- Fetch wait time
- Remote bytes read
- Local blocks read
- Remote blocks read

B.5.2.2 Integrating SQL Metrics

We can further improve the observability by integrating SQL metrics with physical plan through physical operators in application logs. SQL metrics can also be correlated with its physical plan by logging the executed query. The approach helps to troubleshoot and identify production behavior. It has been inspired by the formatted explain mode introduced by Spark 3.0. To use this approach, the Spark OSS repo needs to be extended. Both the physical plan and its SQL metrics need to be logged in the application logs for correlation. In the application, we can access executed query metrics through *KVStore*. We need *SparkListenerSQLExecutionEnd* to retrieve the final values of all metrics per executed query in the driver program. All metrics then can be dumped to application logs. By combining results from the execution end, we can correlate physical operators with the physical plan and their metrics. Please refer to example spark physical with metrics as follows:

```
== Physical Plan ==
+- SortMergeJoin [id#3], [id#11], Inner
   :- Sort [id#3 ASC NULLS FIRST], false, 0
...
== SQL Metrics ==
Sort [CodegenId: 4] [StageIds: 2]
  Metrics: [sort time total (min, med, max): 0 ms (0 ms, 0 ms, 0 ms)
  - peak memory total (min, med, max): 44.5 MB (64.0 KB, 64.0 KB,
    16.1 MB)
  - spill size total (min, med, max): 0.0 B (0.0 B, 0.0 B, 0.0 B)]
SortMergeJoin [CodegenId: 1] [StageIds: 2]
  Metrics: [number of output rows: 2]
...
```

B.5.2.3 The Menu

In this recipe, we saw how to leverage Spark SQL metrics pipeline and how to correlate physical plans with SQL metrics through physical operators in application logs. Logging Spark SQL metrics helps when troubleshooting complex Spark queries, monitoring physical operators' runtime behaviors, and measuring the production workloads in detail. The same approach can also be extended for other Spark runtime metrics by consuming Spark public Rest endpoint.

Ömer Eren Avşaroğulları put a huge effort into making this recipe real. I have been fortunate enough to collaborate on this recipe with him.

B.6 Airbnb Price Prediction

A data scientist can indeed have various responsibilities based on the role definition of a company. Even within a company, there can be different expectations per project, and at that point, your guidance becomes really important to produce a useful working product. To do so, gaining experience, chasing opportunities to improve yourself, and choosing to stay out of your comfort zone get vital to success.

We will give you a data science recipe on a sample project to show what the data science process looks like. Please refer to Figure 8.1 to see the whole process. We will be covering the details step by step. This sample is going to cover many aspects of a data science project, but be aware that this is a sample project. In real-life data science projects, there can be many iterations based on model performance or update on requirements.

B.6.1 Problem Statement

Airbnb is an online marketplace to arrange or offer places to stay for the short term. It practically offers to lodge when you want to visit a place with a taste of the local atmosphere and not compromising the house comfort. Airbnb has mainly two customer profiles: the people who want to stay in a place for a short time and the people who offer places to stay. For our sample project, we are going to focus on the second profile whose one of the problems is the difficulty to decide the right price for a new place since Airbnb does not give a guide on how to choose the relevant price for each place. There is high competition in the online marketplace. If you set up the price to a high value, then you might miss your potential customers, and if the otherwise happens, then you are going to make less profit than you are actually able to do. Hence, we will predict prices for the new houses to guide customers to set up the right price.

B.6.2 Design Approach

After writing down the problem definition, we should check the statement with business owners and ensure that we are focusing on solving the same problem. Then, we should decide on an objective metric to track our improvements to the model. When we think globally, it is not so easy to decide what to measure and improve. As an end goal, we want to guide our customers to define suitable prices for their houses. However, we cannot really define a metric to measure that. So, we should find objective measurable metrics to evaluate our progress. In this recipe, we will use the R2 score since we predict a numerical value where it is the price of a house. This is a regression problem. R2 score is a common metric when looking at regression values. Optimizing a model to have the lowest mean-squared error (MSE) will also optimize a model to have the highest R2 score that is frequently interpreted as the amount of variability captured by a model. Therefore, you can think of MSE as the average amount you miss across all the points and the R2 value as the amount of variability in the points that you capture with a model.

We stated our problem and how to measure the success of our project. It is a good practice to research a problem to check existing papers or solutions to see how other people have approached the problem. It is vital to research when we deal with deep learning techniques.

B.6.2.1 Tech Stack

We will use a python tech stack with Jupyter notebooks. Python is an easy-to-use language with a vast selection of data science and machine learning libraries. It is opensource and still under development with great community support and supports object-oriented, structured, and functional programming. In addition, we can get our prototypes running much faster compared to other programming languages. The following are data libraries selected for this project:

- pandas for data analysis
- numpy for numerical computing
- matplotlib and seaborn for data visualization
- scikit-learn for machine learning

Jupyter notebooks provide an interactive way to work with data, code, implement the model, and document to collaborate with other people. It is recommended to check their documentation and understand what they do and install them if you are going to try this project locally. All the steps described here exist in the Github repository (Kavasoglu, 2019); feel free to follow along with the notebook source code.

B.6.2.2 Data Understanding and Ingestion

Think about the data availability, quality, and suitability for the problem, and brainstorm what datasets can be used to solve the problem. We might use company data alongside public data. For this problem, we will use public data from Airbnb (2020) containing property details, host information, and historical price information. There are some websites depending on the country and city that the statistics per region like sale prices are published. These websites can provide valuable information additionally. Also, we need to be careful about General Data Protection Regulation (GDPR) and not use Personally Identifiable Information (PII)/Protected Health Information (PHI) data unless we have legitimate reasons to use it. We will use the detailed listings data for Istanbul.

B.6.2.3 Data Preparation

The dataset has 19 727 rows and 105 columns. It is important to check each column that you use and understand the values that can carry.

We will drop the columns and rows that have many missing values. We defined the threshold as 80% and removed 10 columns and 0 rows that have over 80% missing values.

We will check each column to see whether we have informative columns or not by using the *value_counts* method like *data.is_business_travel_ready.value_counts()*.

We will drop the columns that have the same value for the entire dataset like *is_business_travel_ready* and the columns that have inconsistent values like market that carry Istanbul, Other (International), and Marmara values that do not make sense together.

Some of the columns do not seem useful at all like URLs, so we will just drop them. Also, some of them might be useful like notes and transit explanations that people might find necessary info in these parts that affects their decisions. However, for simplicity, we will drop them as well. To implement the first working model, we try to think as simple as possible. Simple is better than complicated.

We see that there are some columns with wrong data types like price that should be numerical data type. We will convert them into numerical columns along with Boolean-valued columns. Some columns definitely need extra analysis and processing. We will investigate them in detail under the following groups.

Putting into Rough Buckets Some columns carry very granular data; it is not very helpful for generalization that is the main point of a machine learning algorithm. So, we will merge these values to be distinguishable by rough buckets:

- *host_response_rate*: From 68 unique values into 4 buckets: 0–79%, 80–99%, 100%, and unknown.
- *property_type*: From 40 unique values into 4 buckets: Apartment, Hotel, House, and Other.
- *first_review_diff* and*last_review_diff*: These columns carry NaN values since there are no reviews for some places yet. We cannot fill in these values with a median value or so, because these nonvalues are also meaningful to us. A person can decide to rent a place or not due to the existing reviews. So, we will convert these columns into categorical ones to keep no review info as well. The same goes for review rating columns. When we donot have a review, we donot have a relevant rating measure. We will also convert them into categorical ones for the columns like *reviews_score_communication* and *reviews_score_location*.

Visualizing the Data Drawing histograms can help to decide these buckets because they enable us to see data distribution shown in Figure B.8.

Extracting Useful Information into Separate Columns Amenities carry values as string arrays, e.g. TV, cable TV, and kitchen, and these values are really important when deciding on a place. Thus, we need to extract this data into separate columns to understand the effect of each amenity. We have 154 unique amenities. We will keep 26 ones based on the variety of the data.

Transformation of the Columns Date columns are usually important to understand the time elapse between now and then. So, we will convert these columns into numerical ones to show the day difference between that time and the date when the dataset compiled: *last_review*, *host_since*, and *first_review*.

Removing Outliers Our target variable price has many outlier values. We will remove them from the dataset. Overall in data preprocessing, we should assess

```
objects_processed.first_review_diff.hist(figsize=(10,5),bins=20)
```

```
<matplotlib.axes._subplots.AxesSubplot at 0x1161984a8>
```

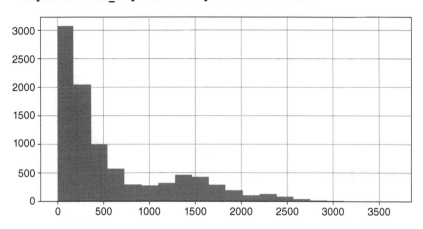

Figure B.8   Airbnb histogram.

how useful the data is and whether we can infer some statistics from this data or not. As a last operation, you can check if there are any correlated columns and remove them if you want to run your models faster. We can use *sns.heatmap* to see correlation scores.

To train and evaluate our model, we will store 70% and 30% of the processed data as train and test datasets, respectively.

B.6.2.4   Modeling and Evaluation

After preprocessing, we have 19 544 rows and 58 columns and 1 target column as price. When we have a look at the statistical distribution of the numerical columns, we can see that some of them have skewed distributions like price, and logarithmic transformation might help to the success of the algorithm. So, we will apply a log transform first to these skewed columns.

Then, we will apply standard scaling for all of the numerical columns and convert categorical columns into numerical ones by applying the *get_dummies* method of pandas.

There are several regression algorithm implementations in *scikit-learn*. We will try out different ones and choose the most successful one according to our objective. When you have the experience, you can choose which ones could work for the dataset and try them in the first place. We will use the training dataset to fit a model and evaluate its performance on our test dataset. For example, we can build a support vector regression (SVR) model like this:

```
svr = SVR()
svr.fit(X_train, y_train)
print('Training score: ', svr.score(X_train, y_train))
y_preds = svr.predict(X_test)
print('Validation score: ', r2_score(y_test,y_preds))
```

When we apply different algorithms like *KNeighborsRegressor*, *GradientBoost-ingRegressor*, *RandomForestRegressor*, etc., we need to think about overfitting and underfitting when evaluating the results. SVR seems to produce meaningful results in our project compared with other algorithms. To improve the results, we can add more feature preprocessing steps, use external datasets like sale prices per region, hyper tune parameters for the algorithm, or try importance sampling on price since it is skewed data. However, it is always a good practice to deploy the first version of our model to see if there are any obstacles to make our model live to minimize the time to the first online experiment. Our first model does not need to be perfect. It can be used as a baseline, and further implementations should be better than the previous ones according to your objective metric.

B.6.2.5 Monitor and Iterate

After deployment, we need to monitor the results and the healthiness of the input dataset. When we build a model, we learn from the statistical distribution of the features. When this distribution changes our model can produce unexpected results as well, it is really important to monitor the statistical distribution of each feature and alert people when there is a significant change, so we can retrain our model when necessary.

We need to collect feedback fast and often to iterate and improve more. You should get feedback from business users, insight managers, or the innovation team along with the metrics you collect from the live product.

Zehra Kavasoğlu put a huge effort into making this recipe real. I have been fortunate enough to collaborate on this recipe with her.

Bibliography

Daniel J. Abadi, Peter A. Boncz, and Stavros Harizopoulos. Column-oriented database systems. *Proceedings of the VLDB Endowment*, 2 (2): 1664–1665, 2009.

Martín Abadi, Paul Barham, Jianmin Chen, Zhifeng Chen, Andy Davis, Jeffrey Dean, Matthieu Devin, Sanjay Ghemawat, Geoffrey Irving, Michael Isard, et al. Tensorflow: A system for large-scale machine learning. In *12th {USENIX} Symposium on Operating Systems Design and Implementation ({OSDI} 16)*, pages 265–283, 2016.

Inside Airbnb. Inside Airbnb, 2020. http://insideairbnb.com/get-the-data.html.

Airflow. Airflow is a platform created by the community to programmatically author, schedule and monitor workflows, 2020. https://airflow.apache.org/.

Saleema Amershi, Andrew Begel, Christian Bird, Robert DeLine, Harald Gall, Ece Kamar, Nachiappan Nagappan, Besmira Nushi, and Thomas Zimmermann. Software engineering for machine learning: a case study. In *2019 IEEE/ACM 41st International Conference on Software Engineering: Software Engineering in Practice (ICSE-SEIP)*, pages 291–300. IEEE, 2019.

James Ang and Thompson S.H. Teo. Management issues in data warehousing: insights from the housing and development board. *Decision Support Systems*, 29 (1): 11–20, July 2000. ISSN 0167-9236. 10.1016/S0167-9236(99)00085-8. http://dx .doi.org/10.1016/S0167-9236(99)00085-8.

Apache Flume. Flume is a distributed, reliable, and available service for efficiently collecting, aggregating, and moving large amounts of streaming event data, 2019. https://flume.apache.org/.

Apache Gobblin. Apache gobblin - a distributed data integration framework, 2020. https://gobblin.apache.org/.

Apache Hadoop. Apache Hadoop, 2006. http://hadoop.apache.org/.

Apache Kafka. Kafka, a distributed streaming platform, 2020. https://kafka.apache .org/25/documentation.html.

Apache Nutch. Apache Nutch, 2004. http://nutch.apache.org/.

Apache PredictionIO. Apache predictionio, 2019. http://predictionio.apache.org/.

Designing Big Data Platforms: How to Use, Deploy, and Maintain Big Data Systems,
First Edition. Yusuf Aytas.
© 2021 John Wiley & Sons, Inc. Published 2021 by John Wiley & Sons, Inc.

Apache Pulsar. Apache pulsar is an open-source distributed pub-sub messaging system, 2020. http://pulsar.apache.org/docs/en/2.5.2/standalone/.

Apache Storm. Apache storm is a free and open source distributed realtime computation system, 2019. https://storm.apache.org/releases/2.1.0/index.html/.

Michael Armbrust, Reynold S. Xin, Cheng Lian, Yin Huai, Davies Liu, Joseph K. Bradley, Xiangrui Meng, Tomer Kaftan, Michael J. Franklin, Ali Ghodsi, et al. Spark SQL: relational data processing in spark. In *Proceedings of the 2015 ACM SIGMOD International Conference on Management of Data*, pages 1383–1394, 2015.

Atlas. Apache atlas – data governance and metadata framework for Hadoop, 2020. https://atlas.apache.org.

Beam. Apache Beam - an advanced unified programming model, 2020. https://beam.apache.org/.

Eric A. Brewer. Towards robust distributed systems. In *PODC*, volume 7, pages 343477–343502. Portland, OR, 2000.

Paulo Caroli. Lean inception, 2017. https://martinfowler.com/articles/lean-inception/.

Cassandra. Apache Cassandra, 2020. https://cassandra.apache.org/.

Michael Cox and David Ellsworth. Managing big data for scientific visualization, 1997.

Benoit Dageville, Thierry Cruanes, Marcin Zukowski, Vadim Antonov, Artin Avanes, Jon Bock, Jonathan Claybaugh, Daniel Engovatov, Martin Hentschel, Jiansheng Huang, et al. The snowflake elastic data warehouse. In *Proceedings of the 2016 International Conference on Management of Data*, pages 215–226, 2016.

DataEthics. Data ethics principles, 2017. https://dataethics.eu/data-ethics-principles/.

Jeffrey Dean and Sanjay Ghemawat. Mapreduce: simplified data processing on large clusters. *Communications of the ACM*, 51 (1): 107–113, January 2004. ISSN 0001-0782. 10.1145/1327452.1327492. http://doi.acm.org/10.1145/1327452.1327492.

Zhamak Dehghani. How to move beyond a monolithic data lake to a distributed data mesh, 2019. https://martinfowler.com/articles/data-monolith-to-mesh.html.

James Dixon. Pentaho, Hadoop, and data lakes. *Blog*, October, 2010.

Bradley Efron. Missing data, imputation, and the bootstrap. *Journal of the American Statistical Association*, 89 (426): 463–475, 1994.

Eric Evans. *Domain-Driven Design: Tackling Complexity in the Heart of Software*. Addison-Wesley Professional, 2004.

Wei Fang, Xue Zhi Wen, Yu Zheng, and Ming Zhou. A survey of big data security and privacy preserving. *IETE Technical Review*, 34 (5): 544–560, 2017.

Apache Flink. Apache Flink is an open source platform for distributed stream and batch data processing, 2020. https://ci.apache.org/projects/flink/flink-docs-release-1.10/.

FluentD. Fluentd is an open source data collector for unified logging layer, 2020. https://docs.fluentd.org/.

Martin Fowler. Event sourcing, 2005. https://www.martinfowler.com/eaaDev/EventSourcing.html.

Nir Friedman, Michal Linial, Iftach Nachman, and Dana Pe'er. Using Bayesian networks to analyze expression data. In *Proceedings of the 4th Annual International Conference on Computational Molecular Biology*, RECOMB '00, pages 127–135, New York, NY, USA, 2000. ACM. ISBN 1-58113-186-0. 10.1145/332306.332355. http://doi.acm.org/10.1145/332306.332355.

Ajit Gaddam. Securing your big data environment. *Black Hat USA*, 2015, 2015.

Alan F. Gates, Olga Natkovich, Shubham Chopra, Pradeep Kamath, Shravan M. Narayanamurthy, Christopher Olston, Benjamin Reed, Santhosh Srinivasan, and Utkarsh Srivastava. Building a high-level dataflow system on top of map-reduce: the pig experience. *Proceedings of the VLDB Endowment*, 2 (2): 1414–1425, 2009.

Mark Grover. Amundsen - Lyft's data discovery & metadata engine, 2019. https://eng.lyft.com/amundsen-lyfts-data-discovery-metadata-engine-62d27254fbb9.

Anurag Gupta, Deepak Agarwal, Derek Tan, Jakub Kulesza, Rahul Pathak, Stefano Stefani, and Vidhya Srinivasan. Amazon redshift and the case for simpler data warehouses. In *Proceedings of the 2015 ACM SIGMOD International Conference on Management of Data*, pages 1917–1923, 2015.

Adam Jacobs. The pathologies of big data. *Communications of the ACM*, 52 (8): 36–44, 2009.

David Karger, Alex Sherman, Andy Berkheimer, Bill Bogstad, Rizwan Dhanidina, Ken Iwamoto, Brian Kim, Luke Matkins, and Yoav Yerushalmi. Web caching with consistent hashing. *Computer Networks*, 31 (11–16): 1203–1213, 1999.

Shachar Kaufman, Saharon Rosset, Claudia Perlich, and Ori Stitelman. Leakage in data mining: formulation, detection, and avoidance. *ACM Transactions on Knowledge Discovery from Data (TKDD)*, 6 (4): 1–21, 2012.

Zehra Kavasoglu. Airbnb Istanbul data playbook, 2019. https://github.com/kavasoglu/airbnb_istanbul.

Vijay Khatri and Carol V. Brown. Designing data governance. *Communications of the ACM*, 53 (1): 148–152, 2010.

Martin Kleppmann. *Making Sense of Stream Processing: The Philosophy Behind Apache Kafka and Scalable Stream Data Platforms*. O'Reilly Media, Inc., 2016.

Martin Kleppmann. *Designing Data-Intensive Applications: The Big Ideas Behind Reliable, Scalable, and Maintainable Systems*. O'Reilly Media, Inc., 2017.

Knox. Knox gateway, 2020. https://knox.apache.org/.

Ron Kohavi et al. A study of cross-validation and bootstrap for accuracy estimation and model selection. In *Ijcai*, volume 14, pages 1137–1145. Montreal, Canada, 1995.

Jay Kreps. Questioning the lambda architecture. *Online article, July*, page 205, 2014. https://www.martinfowler.com/eaaDev/EventSourcing.html.

Patrick Kua, N. Ford, and R. Parsons. *Building Evolutionary Architectures*. O'Reilly Media, Inc., Sebastopol, CA, 2017.

Sanjeev Kulkarni, Nikunj Bhagat, Maosong Fu, Vikas Kedigehalli, Christopher Kellogg, Sailesh Mittal, Jignesh M. Patel, Karthik Ramasamy, and Siddarth Taneja. Twitter Heron: stream processing at scale. In *Proceedings of the 2015 ACM SIGMOD International Conference on Management of Data*, pages 239–250, 2015.

C. Charles Law, William J. Schroeder, Kenneth M. Martin, and Joshua Temkin. A multi-threaded streaming pipeline architecture for large structured data sets. In *Proceedings of the Conference on Visualization '99: Celebrating Ten Years*, VIS '99, pages 225–232, Los Alamitos, CA, USA, 1999. IEEE Computer Society Press. ISBN 0-7803-5897-X. http://dl.acm.org/citation.cfm?id=319351.319378.

Ninghui Li, Tiancheng Li, and Suresh Venkatasubramanian. t-closeness: privacy beyond k-anonymity and l-diversity. In *2007 IEEE 23rd International Conference on Data Engineering*, pages 106–115. IEEE, 2007.

Steve Lohr. The age of big data. *New York Times*, 11 2012.

Ashwin Machanavajjhala, Daniel Kifer, Johannes Gehrke, and Muthuramakrishnan Venkitasubramaniam. l-diversity: privacy beyond k-anonymity. *ACM Transactions on Knowledge Discovery from Data (TKDD)*, 1 (1): 3, 2007.

Ajoy Majumdar and Zhen Li. Metacat: making big data discoverable and meaningful at netflix, 2018. https://netflixtechblog.com/metacat-making-big-data-discoverable-and-meaningful-at-netflix-56fb36a53520.

Nathan Marz. How to beat the cap theorem, 2011. http://nathanmarz.com/blog/how-to-beat-the-cap-theorem.html.

J. R. Mashey. Big data … and the next wave of infrastress. 04 1998.

Viktor Mayer-Schönberger and Kenneth Cukier. *Big Data: A Revolution That Will Transform How We Live, Work, and Think*. Houghton Mifflin Harcourt, 2013.

Sergey Melnik, Andrey Gubarev, Jing Jing Long, Geoffrey Romer, Shiva Shivakumar, Matt Tolton, and Theo Vassilakis. Dremel: interactive analysis of web-scale datasets. *Proceedings of the VLDB Endowment*, 3 (1–2): 330–339, 2010.

Mike Mesnier, Gregory R. Ganger, and Erik Riedel. Object-based storage. *IEEE Communications Magazine*, 41 (8): 84–90, 2003.

Arun C. Murthy, Vinod Kumar Vavilapalli, Doug Eadline, Joseph Niemiec, and Jeff Markham. *Apache Hadoop YARN: Moving Beyond MapReduce and Batch Processing with Apache Hadoop 2*. Addison-Wesley Professional, 1st edition, 2014. ISBN 0321934504, 9780321934505.

Shadi A. Noghabi, Kartik Paramasivam, Yi Pan, Navina Ramesh, Jon Bringhurst, Indranil Gupta, and Roy H. Campbell. Samza: stateful scalable stream processing at linkedin. *Proceedings of the VLDB Endowment*, 10 (12): 1634–1645, 2017.

Mike Olson. HADOOP: scalable, flexible data storage and analysis. *IQT Quart*, 1: 14–18, 2010.

Zeljko Panian. Some practical experiences in data governance. *World Academy of Science, Engineering and Technology*, 62 (1): 939–946, 2010.

Eugenia Politou, Efthimios Alepis, and Constantinos Patsakis. Forgetting personal data and revoking consent under the GDPR: challenges and proposed solutions. *Journal of Cybersecurity*, 4 (1): tyy001, 2018.

Catherine Pope, Susan Halford, Ramine Tinati, and Mark Weal. What's the big fuss about 'big data'? *Journal of Health Services Research & Policy*, 19: 67–68, 2014. 10.1177/1355819614521181.

David Martin Powers. Evaluation: from precision, recall and F-measure to ROC, informedness, markedness and correlation, 2011.

Foster Provost and Tom Fawcett. *Data Science for Business: What You Need to Know About Data Mining and Data-Analytic Thinking*. O'Reilly Media, Inc., 2013.

R. What is R? 2019. https://www.r-project.org/about.html.

Ranger. Apache ranger - introduction, 2020. https://ranger.apache.org/.

Philip Russom et al. Big data analytics. *TDWI Best Practices Report, Fourth Quarter*, 19 (4): 1–34, 2011.

Badrul Sarwar, George Karypis, Joseph Konstan, and John Riedl. Item-based collaborative filtering recommendation algorithms. In *Proceedings of the 10th International Conference on World Wide Web*, pages 285–295, 2001.

Sentry. Apache sentry, 2018. https://sentry.apache.org/.

Raghav Sethi, Martin Traverso, Dain Sundstrom, David Phillips, Wenlei Xie, Yutian Sun, Nezih Yegitbasi, Haozhun Jin, Eric Hwang, Nileema Shingte, et al. Presto: SQL on everything. In *2019 IEEE 35th International Conference on Data Engineering (ICDE)*, pages 1802–1813. IEEE, 2019.

Victoria Stodden. The data science life cycle: a disciplined approach to advancing data science as a science. *Communications of the ACM*, 63 (7): 58–66, 2020.

Latanya Sweeney. Achieving k-anonymity privacy protection using generalization and suppression. *International Journal of Uncertainty, Fuzziness and Knowledge-Based Systems*, 10 (05): 571–588, 2002.

Ashish Thusoo, Joydeep Sen Sarma, Namit Jain, Zheng Shao, Prasad Chakka, Suresh Anthony, Hao Liu, Pete Wyckoff, and Raghotham Murthy. Hive: a warehousing solution over a map-reduce framework. *Proceedings of the VLDB Endowment*, 2 (2): 1626–1629, 2009.

Robin Van Meteren and Maarten Van Someren. Using content-based filtering for recommendation. In *Proceedings of the Machine Learning in the New Information Age: MLnet/ECML2000 Workshop*, volume 30, pages 47–56, 2000.

Ham Vocke. The practical test pyramid, 2018. https://martinfowler.com/articles/practical-test-pyramid.html.

Sholom M. Weiss and Nitin Indurkhya. *Predictive Data Mining: A Practical Guide*. Morgan Kaufmann Publishers Inc., San Francisco, CA, USA, 1998. ISBN 1-55860-403-0.

Rick L. Wilson and Peter A. Rosen. Protecting data through perturbation techniques: the impact on knowledge discovery in databases. *Journal of Database Management (JDM)*, 14 (2): 14–26, 2003.

Longzhi Yang, Jie Li, Noe Elisa, Tom Prickett, and Fei Chao. Towards big data governance in cybersecurity. *Data-Enabled Discovery and Applications*, 3 (1): 10, 2019.

Matei Zaharia, Mosharaf Chowdhury, Michael J. Franklin, Scott Shenker, Ion Stoica, et al. Spark: cluster computing with working sets. *HotCloud*, 10 (10–10): 95, 2010.

Matei Zaharia, Mosharaf Chowdhury, Tathagata Das, Ankur Dave, Justin Ma, Murphy McCauly, Michael J. Franklin, Scott Shenker, and Ion Stoica. Resilient distributed datasets: a fault-tolerant abstraction for in-memory cluster computing. In *Presented as part of the 9th {USENIX} Symposium on Networked Systems Design and Implementation ({NSDI} 12)*, pages 15–28, 2012.

Qing Zheng, Haopeng Chen, Yaguang Wang, Jiangang Duan, and Zhiteng Huang. COSBench: a benchmark tool for cloud object storage services. In *2012 IEEE 5th International Conference on Cloud Computing*, pages 998–999. IEEE, 2012.

Index

Designing Big Data Platforms: How to Use, Deploy, and Maintain Big Data Systems,
First Edition. Yusuf Aytas.
© 2021 John Wiley & Sons, Inc. Published 2021 by John Wiley & Sons, Inc.

Printed and bound by CPI Group (UK) Ltd, Croydon, CR0 4YY